Engelbert Portenkirchner

Charge Transfer onto Redox Mediators

Engelbert Portenkirchner

Charge Transfer onto Redox Mediators

The Charge Transfer from Organic Semiconductors onto Redox Mediators for Electro- and Photochemical CO2 Reduction

Südwestdeutscher Verlag für Hochschulschriften

Impressum / Imprint
Bibliografische Information der Deutschen Nationalbibliothek: Die Deutsche Nationalbibliothek verzeichnet diese Publikation in der Deutschen Nationalbibliografie; detaillierte bibliografische Daten sind im Internet über http://dnb.d-nb.de abrufbar.
Alle in diesem Buch genannten Marken und Produktnamen unterliegen warenzeichen-, marken- oder patentrechtlichem Schutz bzw. sind Warenzeichen oder eingetragene Warenzeichen der jeweiligen Inhaber. Die Wiedergabe von Marken, Produktnamen, Gebrauchsnamen, Handelsnamen, Warenbezeichnungen u.s.w. in diesem Werk berechtigt auch ohne besondere Kennzeichnung nicht zu der Annahme, dass solche Namen im Sinne der Warenzeichen- und Markenschutzgesetzgebung als frei zu betrachten wären und daher von jedermann benutzt werden dürften.

Bibliographic information published by the Deutsche Nationalbibliothek: The Deutsche Nationalbibliothek lists this publication in the Deutsche Nationalbibliografie; detailed bibliographic data are available in the Internet at http://dnb.d-nb.de.
Any brand names and product names mentioned in this book are subject to trademark, brand or patent protection and are trademarks or registered trademarks of their respective holders. The use of brand names, product names, common names, trade names, product descriptions etc. even without a particular marking in this work is in no way to be construed to mean that such names may be regarded as unrestricted in respect of trademark and brand protection legislation and could thus be used by anyone.

Coverbild / Cover image: www.ingimage.com

Verlag / Publisher:
Südwestdeutscher Verlag für Hochschulschriften
ist ein Imprint der / is a trademark of
OmniScriptum GmbH & Co. KG
Heinrich-Böcking-Str. 6-8, 66121 Saarbrücken, Deutschland / Germany
Email: info@svh-verlag.de

Herstellung: siehe letzte Seite /
Printed at: see last page
ISBN: 978-3-8381-3937-1

Zugl. / Approved by: Linz, JKU, Diss., 2014

Copyright © 2014 OmniScriptum GmbH & Co. KG
Alle Rechte vorbehalten. / All rights reserved. Saarbrücken 2014

Abstract

In this work the photoinduced electron transfer from organic semiconductors onto redox mediator catalysts for CO_2 reduction has been investigated. In the beginning, the work focuses on the identification, characterization and test of suitable catalyst materials. For this purpose, rhenium compounds with 2,2'-bipyridine bis(arylimino) acenaphthene ligands and pyridinium were tested for molecular homogenous catalysis.

Infrared, ultraviolet-visible (UV-Vis) and nuclear magnetic resonance (NMR) spectroscopy were used for initial characterization of the catalyst substances. Since the interpretation of infrared spectra was difficult for large molecules based on measured data only, additionally infrared absorption spectra obtained by quantum mechanical density functional theory (DFT) calculations were successfully used to correlate characteristic features in the measured spectra to their molecular origin. It was found that experimentally observed data and quantum chemical predictions for the infrared spectra of the novel compounds are in good agreement. Additionally, quantum mechanical calculations were carried out for the determination of molecular orbital frontier energy levels and correlated to UV-Vis absorption and cyclic voltammetry measurements.

Extensive cyclic voltammetry measurements and bulk controlled-potential electrolysis experiments were performed using a N_2- and CO_2-saturated electrolyte solution. Together with a detailed product analysis via infrared spectroscopy, gas and ion chromatography the results allowed electrochemical characterizations of the novel catalysts regarding their suitability for electrochemical CO_2 reduction.

Once suitable catalysts were identified, the materials were immobilized on the electrode surface by electro-polymerization of the catalyst (5,5'-bisphenylethynyl-

2,2'-bipyridyl)Re(CO)$_3$Cl itself or by incorporation of (2,2'-bipyridyl)Re(CO)$_3$Cl into a polypyrrole matrix, thereby changing from homogeneous to heterogeneous catalysis. In an entirely new approach the catalyst (2,2'-bipyridyl)Re(CO)$_3$Cl was covalently attached to poly(3-hexylthiophene).

Finally, the combination of suitable catalysts with organic semiconductor materials was investigated for photoinduced energy and charge transfer from the donor semiconductor polymer to the catalyst acceptor. Poly(N-vinylcarbazole) (PVK) was used as absorber material acting as efficient redox photosensitizer in combination with (2,2'-bipyridyl)Re(CO)$_3$Cl as catalyst acceptor.

Photoluminescence quenching experiments in solid film mixtures and in a solid-liquid interface between polymer and catalyst revealed strong luminescence quenching by the catalyst material due to a resonant energy transfer.

This work shows, that the combination of organic semiconductors with catalyst redox mediators offers a promising approach for electro- and photocatalytic CO_2 reduction.

Kurzfassung

Die vorliegende Arbeit beschäftigt sich mit dem photoinduzierten Ladungstransfer zwischen organischen Halbleitern und Katalysator-Redox-Mediatoren mit dem Ziel der chemischen Reduktion von CO_2. Geeignete Katalysator- Moleküle konnten identifiziert und charakterisiert werden. Anhand vielversprechender Ergebnisse wurden Rhenium enthaltende Übergansmetallkomplexe mit 2,2'-bipyridin und bis(arylimino) acenaphten Liganden genauer auf deren Eigenschaften zur homogenen Katalyse getestet.

Für die anfängliche Charakterisierung der Katalysatoren wurden Infrarot-, UV-Vis- und Kern-Resonanz- (NMR)-Spektroskopie Messungen durchgeführt. Weiters wurden Infrarotabsorptionsspektren anhand quantenmechanischer Berechnungen mittels Dichte-Funktional-Theorie (DFT) erfolgreich berechnet und charakteristische Absorptionsbanden konnten den entsprechenden Molekülschwingungen zugeordnet werden. Die gemessenen und berechneten Spektren weisen eine gute Übereinstimmung auf.

Zusätzlich wurden quantenmechanische Berechnungen für die Bestimmung der jeweiligen Molekülorbitale verwendet und die Ergebnisse der Berechnungen mittels experimentellen UV-Vis- und Cyclovoltammetrie-Messungen verglichen. Intensive Cyclovoltammetrie- und Elektrolyseexperimente in N_2- und CO_2- gesättigten Elektrolytlösungen wurden in Zusammenhang mit einer detaillierten Analyse der Reduktionsprodukte durchgeführt. Hierfür wurden Infrarotspektroskopie, Gas- und Ionenchromatographie verwendet. Die neuen Katalysatoren konnten hinsichtlich deren potentiellen Verwendbarkeit zur CO_2 Reduktion somit detailliert charakterisiert werden.

Nachdem passende Katalysatoren gefunden worden waren, konnten diese auf der Elektrodenoberfläche durch Elektropolymerisation des Katalysators (5,5'-

bisphenylethynyl-2,2'-bipyridyl) $Re(CO)_3Cl$ selbst oder durch Einbettung von (2,2'-bipyridyl)$Re(CO)_3Cl$ in eine Polypyrrol-Matrix immobilisiert werden. Dies ermöglichte den Übergang von einer homogenen zu einer heterogenen Katalyse. In einem komplett neuen Ansatz wurde der Katalysator (2,2'-bipyridyl) $Re(CO)_3Cl$ kovalent an Poly(3-Hexylthiophen) gebunden.

Abschließend wurde der photoinduzierte Energie- und Ladungstransfer durch Kombination eines Polymerhalbleiters als Donor und eines Katalysators als Akzeptor untersucht. Poly(N-vinylcarbazol) (PVK) als effektiver Photosensibilisator wurde in Kombination mit dem als Akzeptor fungierendem (2,2'-bipyridyl) $Re(CO)_3Cl$ getestet.

Experimente zur Löschung der Photolumineszenz in Feststoffmischungen und in Fest-Flüssig-Grenzschichten zwischen Polymer und Katalysator zeigten eine starke Lumineszenz-Löschung aufgrund eines resonanten Energietransfers.

Zusammenfassend konnte in dieser Arbeit gezeigt werden, dass die Kombination von organischen Halbleitern mit Katalysator-Redox-Mediatoren eine aussichtsreiche Methode zur elektro- und photokatalytischen CO_2-Reduktion darstellt.

Contents

1 Introduction   9

    1.1 Energy and environment 9

    1.2 The end of cheap fossil fuels 11

    1.3 Carbon capture and utilization 13

    1.4 Organic semiconductors for CO_2 reduction 17

2 Experimental techniques   21

    2.1 Electrochemistry 21

        2.1.1 Electrochemistry fundamentals 21

        2.1.2 The electrical double layer 25

        2.1.3 Cyclic voltammetry experiments 27

        2.1.4 Controlled potential electrolysis 29

        2.1.5 Electropolymerisation 34

    2.2 Spectroscopic methods 36

        2.2.1 UV-Vis absorption and PL spectroscopy 36

        2.2.2 FTIR absorption spectroscopy 38

		2.2.3 ATR-FTIR absorption spectroscopy	42
	2.3	Analytical methods for product analysis	43
		2.3.1 Gas chromatography	44
		2.3.2 Ion chromatography	46
	2.4	Homogeneous photocatalysis	46
	2.5	Catalyst materials	47
3	**On the nature of rhenium-(I) bipyridine complexes**		**49**
	3.1	Symmetry and structure	49
	3.2	The photostability of rhenium-(I) bipyridine complexes	52
	3.3	The Jablonski-Diagramm for rhenium-(I) bipyridine complexes .	54
4	**Photophysical results**		**57**
5	**Quantum chemical calculations**		**61**
	5.1	Infrared absorption spectra	62
	5.2	Molecular orbital energy levels	64
6	**Homogeneous electro catalysis**		**69**
	6.1	Rhenium compounds with 2,2'-bipyridine ligands	70
	6.2	Rhenium compounds with bis (arylimino) acenaphthene	83
	6.3	Pyridinium and pyridazinium as catalyst	91
7	**Homogeneous photo catalysis**		**105**

7.1	Results on rhenium-(I) bipyridine complexes	105
7.2	Comparing photo- and electrochemistry	109

8 Heterogeneous electro catalysis 113

8.1	(2,2'-bipy.)Re(CO)$_3$Cl incorporation into a polymer matrix	113
8.2	(5,5'-bisphen.-2,2'-bipy.)Re(CO)$_3$Cl polymerization	117
8.3	(4,4'-dicarboxyl-2,2'-bipy.)Re(CO)$_3$Cl immobilization	128

9 Organic semiconductors for CO_2 reduction 131

9.1	Semiconductor – electrolyte interface		132
9.2	Poly(3-hexylthiophene) electrodes with pyridinium		137
9.3	Polyvinylcarbazole with (2,2'-bipy.)Re(CO)$_3$Cl		143
	9.3.1	Photoluminescence quenching in solid films	145
	9.3.2	Photoluminescence quenching in ACN solution	151
9.4	Functionalized Poly(3-hexylthiophene)		153
	9.4.1	Electropolymerisation and characterization	155
	9.4.2	Cyclic voltammetry studies on electrochemical CO_2 reduction	159

10 Summary and future studies 163

10.1	What has been accomplished	163
10.2	The need of novel catalysts	169
10.3	A more elaborate view on CO_2 reduction to date	171

11 Appendix **177**

Chapter 1

Introduction

1.1 Energy and environment

The current energy supply of human society is mainly based on fossil fuels like coal, oil and gas which are related to several problems. The reserves are decreasing and their final depletion seems to be just a matter of time.[1] In addition, combustion of fossil fuels leads to the emission of CO_2, which is considered as a main greenhouse gas.[2] Figure 1.1(a) and 1.1(b) show measurements of the atmospheric carbon dioxide concentration at Mauna Loa, Hawaii. The presented data reveals an increase from initially 300 parts per million (ppm) to more than 400 ppm current level over the last decades.[3] Figure 1.1(b) shows a detail monthly view of the data represented in Figure 1.1(a) where the annual fluctuations in CO_2 concentration according to the seasonal vegetation changes in the northern and southern hemispheres are apparent. The possible consequences of the man made greenhouse effect (increased temperatures) and the depletion of fossil fuels on oncoming generations depict that there is a need for environmentally friendly and sustainable energy systems.[4]

Several approaches have been made to replace fossil fuels by sustainable techniques like solar and wind energy, hydroelectric power, energy from biomass, etc. Up to now none of these were able to fully replace fossil fuels. From all available renewable energy resources, solar energy will be the only known source that provides sufficient energy to meet the increasing energy demand of humanity. The total solar energy absorbed by Earth's atmosphere, oceans and land masses is approximately 3,850,000 exajoules (EJ) per year. Even

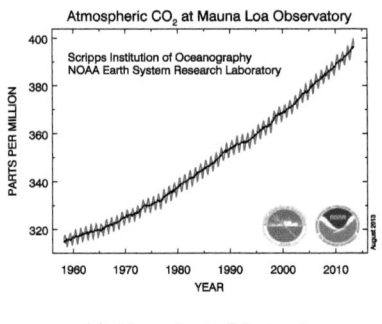
(a) Atmospheric CO_2 level

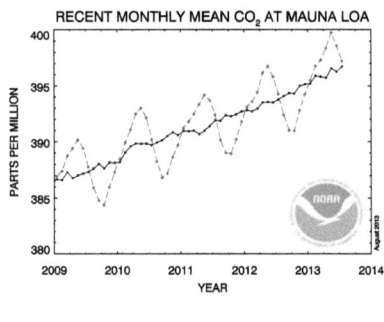
(b) Monthly mean CO_2 level

Figure 1.1: (a) Atmospheric carbon dioxide concentration measured at Mauna Loa, Hawaii over the last decades. (b) Detail monthly view of the data represented in Figure (a) showing the atmospheric carbon dioxide concentration and its seasonal fluctuations as measured at Mauna Loa, Hawaii for the last four years.[3]

with photovoltaic technology available today only 1% of the area of global deserts would be sufficient to produce the entire annual human primary energy consumption as electric power.[5]

Renewable energy sources, like solar radiation, are most abundant in remote locations and have huge fluctuations over short (daily variations) and long timescales (seasonal changes). Therefore cost effective energy storage and transport is just as important as energy generation. This could in theory be achieved by the use of battery systems, but at present batteries are far too expensive and transport of electricity over long distances via cables involves significant losses.[6]

One of the most efficient ways to store energy regarding to energy density and transportability is, as nature has demonstrated over millions of years, chemical energy storage in hydrocarbons. The energy density of gasoline is around 46 MJ/kg (12.8 kWh/kg), which is significantly higher compared to modern batteries, like for example a Li-ion battery, with an energy density of around 540 kJ/kg (150 Wh/kg).[7] Therefore, direct and efficient conversion of solar energy to hydrocarbons would be a major breakthrough on the way towards a sustainable energy supply for humanity.

Figure 1.2 illustrates the idea of energy as a vector in space and time, where

CHAPTER 1. INTRODUCTION

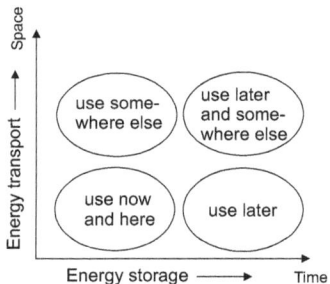

Figure 1.2: The schematic representation of energy as a vector in space and time.

both components have to be fulfilled to meet human demands now and in the future.

1.2 The end of cheap fossil fuels

The human demand for energy is growing steadily. This is anticipated to be even more pressing since the global population and economies are increasing rapidly too. Succeeding current trends it is predicable that the world energy demand is following. During the last 25 years alone, world energy consumption has increased by about 60 %.[8]

Oil is still the most important energy source for mankind. This becomes apparent if we take into account that each US citizen for example is consuming about 11 liter of oil per person per day.[9] Taking global oil consumption into account, experts expect an increase of about 50% by 2030. This will become progressively problematic since the easiest oil to extract has by now been produced and new reserves are difficult to find.[1] Humanity has tried to keep pace with increasing consumption, but the rapid increase in crude oil extraction by the modern industrialized society within the last century is anticipated to come to a maximum in the near future. This point at which global oil production will reach a final peak and then decline is well known in literature as "peak oil" and has been around for decades. Some scientists argue that this point in time has already passed and oil production will decline in the near future. Although statistics show that oil reserves are apparently increasing, the total percentage which is available for extraction is steadily decreasing. If the United States, as

1.2. THE END OF CHEAP FOSSIL FUELS

the biggest oil consumer in the world, are taken as an example, production as a percentage of reserves has declined from 9 % in 1980 to 6 % by 2012. On a global scale the production at existing oil fields is continuously declining at rates of about 4.5 % to 6.7 % per year.[10, 11, 12]

Taking this data into account leads to the conclusion that although humanity will not run out of oil soon, but of oil that can be produced easily and cheaply. Figure 1.3 shows the oil production from 1998 to 2011 and the corresponding Brent crude oil price in US dollar per barrel. It can be seen that the oil production followed the oil demand until 2005 and then leveled off although the oil price continued to increase.[10, 13]

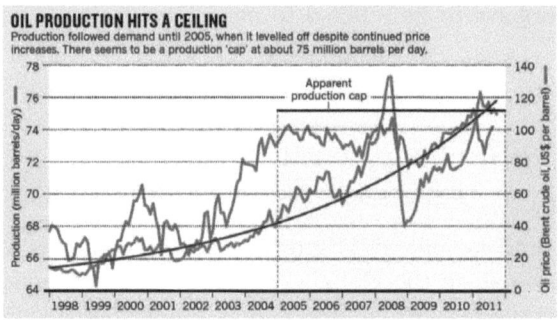

Figure 1.3: Oil production from 1998 to 2011 and corresponding Brent crude oil price in US dollar per barrel. The oil production followed the oil demand until 2005 and then leveled off although the oil price continued to increase. (Graph reproduced from reference [10] and the data therein is taken from reference [13].)

Although the US Energy Information Administration optimistically planed with a 30 % increase in oil production between now and 2030, it is stated that all of that increase is in the form of unidentified projects. This unidentified projects is oil yet to be discovered.[13] Murray and King pointed out in their article of 2012,[10] that even if production at existing fields miraculously stopped declining, such an increase would require 22 million barrels per day of new oil production by 2030. Taking however current declines of 5 % per year into account they calculated that this would need a discovery and utilization of new fields yielding more than 64 million barrels per day. This amount is equivalent to todays total oil production. The authors further concluded that this is very unlikely to happen.

It is clear that the scarcity of oil will heavily affect the global economy,

CHAPTER 1. INTRODUCTION

which is so closely tied to physical resources and a rapidly growing world economy will require large increases in energy supplies over the next few years. Following this argumentation it is interesting and important to notice that the approaches needed for tackling the economic impacts of resource scarcity and climate change described in the previous chapter are the same and obvious. Human society has to move away from the dependence on fossil-fuel energy sources. Learning a lesson from history it tells that the implications of climate change have driven only slow policy responses. On the other hand economic consequences tend to drive shorter-term action. It is known from past records that when there are oil price spikes, the economy begins to respond within a year or less.

A solution to this multiple problem space, whether it is climate or economy driven, can only be done by increasing efficiency of transportation, residential, commercial, and industrial uses and thereby decouple energy consumption and economic growth.[8] Furthermore global energy supply has to move away from fossil fuels like coal, oil and gas to renewable energy sources. To make this transition possible, the capture and utilization of carbon dioxide from earths atmosphere and its transformation to a carbon neutral energy carrier, is of utmost importance.

1.3 Carbon capture and utilization

Possible processes for carbon dioxide (CO_2) recycling were formerly described in the literature and include the capture of CO_2 from air [14], release of pure CO_2 and finally fuel synthesis.[15, 16] For fuel synthesis CO_2 has to be chemically reduced. This can be done either directly (electrochemically or photochemically) or in an indirect way using for example hydrogen gas as a reducing agent. In both methods a catalyst is usually needed. Whereas the indirect reduction with hydrogen is a well-known process, the direct reduction methods are still a topic of basic research and also the focus of this thesis.

CO_2 reduction processes have already been implemented with good efficiencies through various strategies. Two interesting approaches were outlined by Stucki et. al [15] and Specht et. al. [16] Both attempts used metal hydroxide based solutions, either a calcium hydroxide ($Ca(OH)_2$) or a alkali hydroxide

(NaOH, KOH) solution for CO_2 capturing in an absorption liquid and acidification methods for the CO_2 release. The pure CO_2 gas has then to be reduced with the aid of a reducing agent, here hydrogen gas, in a high temperature thermochemical reactor. The main disadvantages of this approach are the high energy and engineering intensive process steps. As proposed by Olah [17], another way to reduce (CO_2) would be the use of a regenerative electrochemical cell system based on the fuel cell concept. Such a regenerative fuel cell would be able to operate in two modes, namely forward and reverse. In forward mode, the fuel cell generates electricity by oxidizing hydrocarbon fuels such as methane or methanol. In reverse mode, the fuel cell produces these hydrocarbon fuels by the reduction of CO_2. Although the fundamental principles of these two processes are known, no efficient selective reduction of CO_2 to higher hydrocarbons has been achieved by now. Furthermore this method will not answer the question where the electricity for fuel production should come from.

Methods for direct electrochemical activation of carbon dioxide have long been studied and significant progress has been made over the past decade.[18, 19, 20, 21] The electrochemical reduction of CO_2 usually requires a high potential of about -1.9 Vvs. NHE for a one electron activation process. This high potential is partly explained by the necessary large inner-shell rearrangements upon electron transfer and the corresponding bending of the molecule upon one electron activation resulting in an O-C-O angle of 133°.[22, 23] By performing the CO_2 reduction in a multi-electron process, however, this transformation can be facilitated. As a consequence of this multi-electron and proton transfer, the required reduction potentials are lowered significantly. This is reflected by the change in the equilibrium redox potentials between reaction 1.1 and 1.2 where the redox potential changes from -1.90 V to -0.65 V vs. NHE respectively.[24] Furthermore the equilibrium potential of the CO_2 reduction progressively shifts to more positive values as the number of electrons and protons participating in the reduction process increase, comparing different net reactions 1.3 to 1.8.

CHAPTER 1. INTRODUCTION

$$CO_2 + e^- \rightleftharpoons CO_2^{\bullet -} \tag{1.1}$$

$$2CO_2 + 2e^- \rightleftharpoons CO + CO_3^{2-} \tag{1.2}$$

$$CO_2 + 2H^+ + 2e^- \rightleftharpoons HCOOH \tag{1.3}$$

$$CO_2 + 2H^+ + 2e^- \rightleftharpoons CO + H_2O \tag{1.4}$$

$$CO_2 + 4H^+ + 4e^- \rightleftharpoons C + 2H_2O \tag{1.5}$$

$$CO_2 + 4H^+ + 4e^- \rightleftharpoons HCHO + H_2O \tag{1.6}$$

$$CO_2 + 6H^+ + 6e^- \rightleftharpoons CH_3OH + H_2O \tag{1.7}$$

$$CO_2 + 8H^+ + 8e^- \rightleftharpoons CH_4 + 2H_2O \tag{1.8}$$

Although the equilibrium reduction potentials for a proton coupled multi-electron CO_2 reduction appear to be tolerably low, in reality however, the actual reduction potentials on bare metal electrodes are much higher than the Nernst potential due to barrier induced over potentials. In order to decrease the actual redox potential needed for the CO_2-reduction process, suitable catalysts are required.

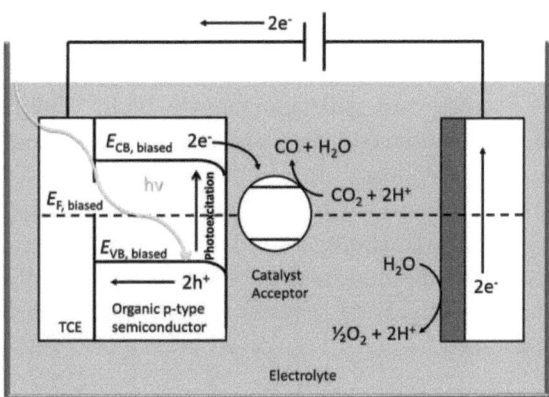

Figure 1.4: The schematic proposal of mimicking photosynthesis by providing a photo induced electron transfer from an organic semiconductor onto a redox mediator catalyst for CO_2 reduction in an electrolyte containing system.

An interesting achievement in this direction was reported by Delacourt et. al.[25], who described syngas conversion into methanol using an electrochemical cell similar to PEM fuel cells. This group achieved 45 % energy efficiency at

$10\,\text{mA cm}^{-2}$, but unfortunately the efficiency was decreasing with higher current densities (30 % at $100\,\text{mA cm}^{-2}$). The homogeneous catalysis of carbon dioxide assimilating fuel generation based on water splitting represents another interesting approach in solar fuel research.[26, 27] These artificial photosynthetic systems are inspired by the overall function of the natural enzymatic reactions, where the industrially important role of hydrogen gas as a two-electron reductant for CO_2-fixation is replaced by organic redox cofactors such as NADH, acting as hydride transfer reagents. In all kinds of technical scenarios following this direction, this leads to the fundamental problem of catalytic cofactor regeneration, i. e. a reversible NAD^+ to NADH conversion to supply electrons and protons for a multistep carbon dioxide conversion into energy rich compounds.[28, 29, 30]

At present most of the best studied catalysts are metal complexes with bipyridine ligands, where the catalyst center consists of transition metals based on rhenium (Re), rhodium (Rh) or ruthenium (Ru). Despite their high current efficiencies and high selectivity, problems in the field of artificial solar fuel production by these catalysts are manifold. Although these molecular catalyst compounds can be used to stabilize intermediate steps of the CO_2 reduction process and thus lower the required overpotential, achieving a simultaneous multiple electron and proton transfer as indicated by the described reactions 1.3 to 1.8 is kinetically extremely difficult to realize and over potentials of most reported catalyst systems are still significantly high. Furthermore, systems based on these catalyst materials, as reported up to now, suffer from low stability and low turnover frequency. In recent yeas Bocarsly et. al. [31] reported on the catalytic reduction of CO_2 to methonol by Pyridinium (the protonated form of Pyridine). Although this system demonstrates only a very low rate of reaction due to a complicated mechanism, which is not jet fully understood, the approach seems to very promising regarding earth abundant catalyst materials and proton rich products.

In the presented thesis the aim is to investigate a new approach for the reduction of carbon dioxide (CO_2) to various desirable products. The intention is to establish a system providing a photo induced electron transfer from an organic semiconductor onto a redox mediator for CO_2 reduction. This can be achieved by using an organic semiconducting material, as it is known today in the field of organic photovoltaic, in contact with an electrolyte solution that

CHAPTER 1. INTRODUCTION

contains a catalyst, which is specific for a multi electron CO_2 reduction. Such a system would mimic photosynthesis which is demonstrated successfully by nature over million of years. In Figure 1.4 a schematic proposal for this process is shown.

Although several steps will be investigated and described in detail, which are necessary to achieve such a system, the main part of this thesis will focus on catalyst research for CO_2 reduction. Once the right materials are specified, spectroelectrochemical methods, such as in situ Fourier transform infrared spectroscopy (FTIR) and current density-voltage characteristics of the catalyst or polymer electrolyte interface, will be investigated to describe the physics behind this processes. At the end a brief outlook on future challenges with respect to CO_2 reduction is presented, together with possible suggestions how this problems could be addressed by the scientific community. In a more elaborate view on CO_2 reduction to date (compare chapter 10) current achievements in the field of direct CO_2 reduction are brought into context of energy supply and related costs.

1.4 Organic semiconductors for CO_2 reduction

A very promising approach in the last decade was the combination of catalysts with inorganic semiconductor materials. In a recent work Kubiak et. al. showed light-assisted co-generation of CO and H_2 from CO_2 and H_2O with a p-type silicon working electrode. The experiments were preformed in an acetonitrile/water mixture, exhibiting high Faradic efficiency at low homogeneous catalyst concentration.[19, 32] Bocarsly and coworkers [33] reported selective reduction of CO_2 to methanol by combining p-GaP semiconductor electrode and pyridine as catalyst material. This attempt of a kinetically difficult $6e^-$ aqueous photo reduction of CO_2 to methanol allows astonishingly low reduction potentials below the standard reduction potential of $-0.52\,\mathrm{V}$ vs SCE.

In this work the approach is different in a way that an organic semiconductor is used as light absorbing donor material in combination with an acceptor catalyst, as for example Pyridinium or $(2,2\text{'-bipyridyl})Re(CO)_3Cl$, for the actual CO_2 reduction. The advantage lies in the unique properties of organic semiconductors as they are flexible, light weight, bio-degradable and bio-compatible,

1.4. ORGANIC SEMICONDUCTORS FOR CO_2 REDUCTION

abundant and hold hence the promising perspective to be cheap in production. Figure 1.5 shows the general idea of a photoinduced charge transfer from a biased organic semiconductor onto a catalyst redox mediator for CO_2 reduction.

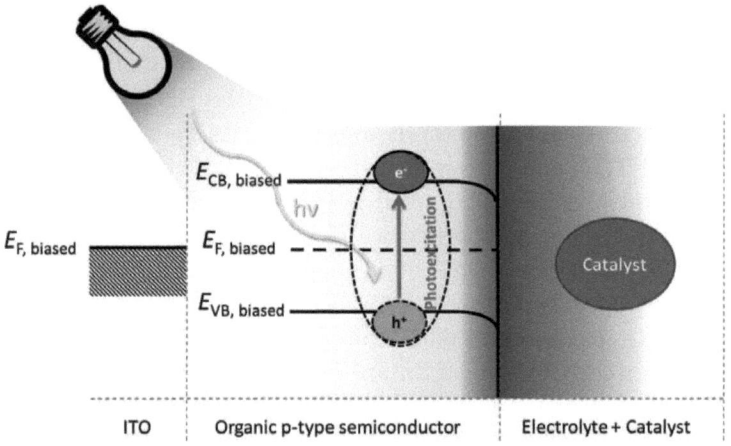

Figure 1.5: Suggested mechanism for photoinduced charge transfer from a biased organic p-type semiconductor onto a catalyst redox mediator for CO_2 reduction.

The schematic drawing in Figure 1.5 shows an organic p-type semiconductor, as for example poly(3-hexylthiophene) (P3HT) on indium tin oxide (ITO), which serves as a transparent conducting electrode (TCE). The whole system is in contact with an electrolyte solution forming a Schottky type of contact.[34] The electrolyte solution contains a catalyst acceptor molecule as for example pyridinium or (2,2'-bipyridyl)Re(CO)$_3$Cl. Upon light irradiation an electron-hole pair is generated in the organic semiconductor. The electron-hole pair generated can not be treated as individual charge carriers since the positive hole (h^+) and the negative electron (e^-) are initially still bound in the material by their Coulomb attraction.[35] The electron-hole pair, also known as exciton, moves as a quasi particle in the bulk of the material until the two charge carriers recombine or hit an interface where the driving force is strong enough for both charges to separate. In the ideal case the exciton will reach the semiconductor-electrolyte interface where the electron transfers to the catalyst acceptor molecule and the hole, now free to move in the bulk material, travels back to the biased ITO contact where it recombines with an electron. The catalyst material loaded with the negative charge is then capable of transferring this charge to the substrate, namely CO_2, and reducing it to the desired product.

Chapter 2

Experimental techniques

2.1 Electrochemistry

2.1.1 Electrochemistry fundamentals

In electrochemical measurement usually a current is flowing through the system. If a current flows, the system is no longer at equilibrium. To gain kinetic information one has to move away from equilibrium conditions but investigate the formation of the equilibrium in the electrochemical setup. For a redox reaction where a species D is reduced to species D$^-$ one can write the reaction according to equation 2.1.

$$D + e^- \xrightarrow{k_{red}} D^- \tag{2.1}$$

The effect then of an applied voltage on the change of the free energy ΔG will follow to the first approximation a linear relationship according to equation 2.2.

$$\Delta G'_{red} = \Delta G_{red\ no\ voltage} + \alpha F V \tag{2.2}$$

Where α is the transfer coefficient (typically 0.5), F is the Faraday constant and V is the applied voltage. This expression can then be substituted into the

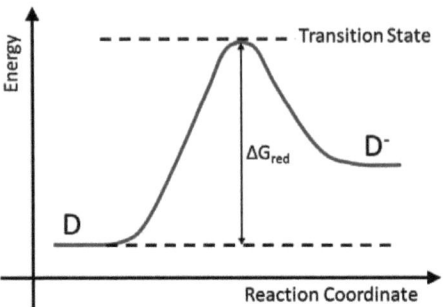

Figure 2.1: Schematic representation of the energy required for the reduction of D to D⁻ going over an energy intensive transition state.

Arrhenius rate equation 7.3 which will result in equation 2.3.

$$k_{red} = A \cdot e^{\left(\frac{-\Delta G_{red\,no\,voltage}}{RT}\right)} \left(\frac{-\alpha FV}{RT}\right) \qquad (2.3)$$

So the rate constant for electron transfer is, to the first approximation, proportional to the exponential of the applied potential. The current in any electrochemical cell is determined by the mass transport to the electrode and the electron transfer process at the surface. This combination gives rise to a current maximum as is typically observed in cyclic voltammetry measurements. This can be understood by the fact that after the potential is reached at which electron transfer is started, the surface concentration of the reacting species falls as the potential increases further, compare Figure 2.2.

The rate for the electron transfer is typically very fast compared to the rate of the mass transport. Heterogeneous charge transfer rate constants (k) are typically around 0.01 to 6 cm s⁻¹, while diffusion coefficients (D) are around 10^{-9} to 10^{-5} cm² s⁻¹.[36, 37] So in practice it is the mass transport to the electrode, consisting of diffusion, convection and migration, that is the limiting step in electrochemical systems. If the system is not stirred or heated the most important contribution to the mass transport comes from diffusion and can be understood by the use of Fick's laws, compare equation 2.4 and 2.5.

CHAPTER 2. EXPERIMENTAL TECHNIQUES

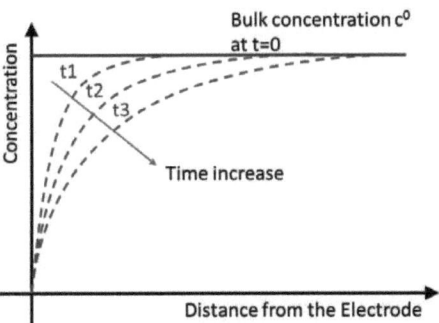

Figure 2.2: Schematic representation of the changing surface concentration of the reacting species after the electron transfer is started.

$$J = -D\frac{\partial c}{\partial x} \qquad (2.4)$$

$$\frac{\partial c}{\partial t} = -\frac{\partial J}{\partial x} \qquad (2.5)$$

Where J is the particle current density, D is the diffusion coefficient in cm^2 s^{-1}, according to the Stokes-Einstein equation 2.6 , and c is the concentration.

$$D = \frac{k_B T}{6\pi \eta R_0} \qquad (2.6)$$

Where η is the dynamic viscosity of the solvent, R_0 is hydrodynamic radius of the diffusing particles and the other symbols have their usual meaning. If now the J from equation 2.4 is substituted into equation 2.5, one gets an equation for the diffusion in relation to the change in concentration over time according to equation 2.7:

$$\frac{\partial c}{\partial t} = D\left(\frac{\partial^2 c}{\partial x^2}\right) \qquad (2.7)$$

This is a partial differential equation of 2nd order and it is in its form very similar to the heat equation derived from Fourier's law of conduction.[38,

p. 227-231] Solving then equation 2.7 under the boundary conditions of equation 2.8:

$$j = -nFD\left(\frac{\partial c}{\partial x}\right)_0 \qquad (2.8)$$

and that at the beginning (at t=0), the concentration is the bulk concentration of the solution c^0 and at any times during the experiment (so at t¿0) the concentration is the electrode surface concentration c^s, will eventually result in equation 2.9.

$$j = -nF\left(\frac{D}{\pi}\right)^{1/2}\frac{c^0 - c^s}{\sqrt{t}} \qquad (2.9)$$

A schematic plot of current over time can be seen in Figure 2.3. When the same data is plotted as $1/\text{time}^{1/2}$, the data follow a straight line where the slope of the line allows calculating D.[38, p. 227-231]

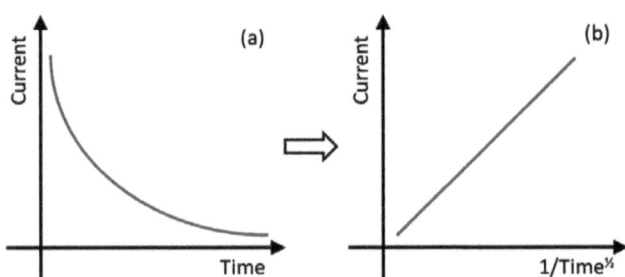

Figure 2.3: (a) Schematic representation of the typical behavior of electrochemical current over time data according to equation 2.9. (b) When the same data is plotted as $1/\text{time}^{1/2}$ the data follow a straight line.

Such a plot for a real experiment is depicted in Figure 2.9 in section 2.1, where typical current density vs. time (a) and $1/\text{time}^{1/2}$ (b) is plotted for the potentionstatic CO_2-electrolysis experiment of Re(5,5'-bisphenylethynyl-2,2'bipyridyl)$(CO)_3$Cl (1-3) at constant $-1950\,\text{mV}$ vs.NHE.

One has to take into account that the above equation is not valid at extremely short time scales. The reason is, that at very short times the concen-

CHAPTER 2. EXPERIMENTAL TECHNIQUES

tration profile, as depicted in Figure 2.2, will be very steep and it is physically not possible for the current to be limited initially by diffusion. In the early times, the electron transfer process will be limiting, but the time to establish a regime where diffusion is limiting will take a fraction of a second.

The cathodic current for any reduction reaction can be expressed by equation 2.10.

$$i_{cathodic} = -nFAk_{red}[conc] \qquad (2.10)$$

Where k_{red} is determined by equation 2.3, and the Nernst equation then predicts the relation between concentration and voltage in equilibrium according to equation 2.11.

$$E = E^0 + \frac{RT}{nF} \ln \frac{[ox]}{[red]} \qquad (2.11)$$

Where E is the applied potential and E^0 is the standard electrode potential and $[ox]$ and $[red]$ are the concentrations of the oxidized and reduced species respectively. For cyclcovoltammetry measurements the rate of electron transfer is typically fast in comparison to the scan rate. Therefore at the electrode surface, equilibrium is established identical to that predicted by thermodynamics. For sufficiently high scan rates however, one observes a shift in the peak potential at a certain, critical scan rate v_c. A plot of peak potential vs. the logarithm of the scan rate can then be used to calculate the heterogeneous electron transfer rate constant k_e For details compare reference nr. [36].

2.1.2 The electrical double layer

The interaction between the ions in the electrolyte solution and the electrode surface are assumed to be of electrostatic origin and result from the fact that the electrode holds an access of (or a deficiency of) electrons at the surface. In order for the interface to remain neutral the charge held on the electrodes is balanced by a redistribution of the ions close to the electrode surface.

2.1. ELECTROCHEMISTRY

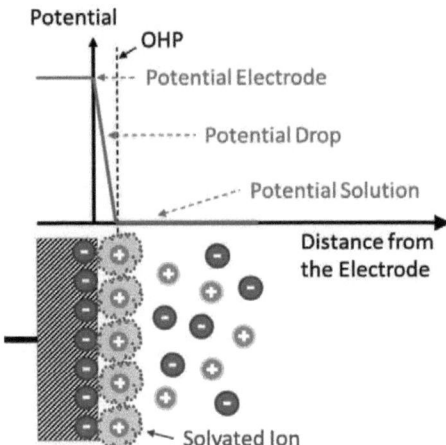

Figure 2.4: Schematic representation of two layers of charge (the double layer) forming at the interface between the electrode and the electrolyte. The potential drop is confined to this double layer region (also called the outer Helmholtz Plane, OHP) in solution.

The distance of approach of the ions is assumed to be limited to the radius of the ion and a single sphere of solvation around each ion. This results in an overall two layers of charge (e.g. double layer) and a potential drop to only this region also known as outer Helmholtz Plane (OHP) . This model is analogous to the potential distribution in an electrical capacitor with two plates of opposite charge separated by some distance d and the potential drop being linear between the two plates. The capacitance (C) of the electrical double layer can then be approximated by equation 2.12.

$$C = \frac{A}{d} \epsilon_0 \epsilon_S \qquad (2.12)$$

Where A is the electrode surface area, d is the distance from the electrode surface to the outer Helmholtz Plane and ϵ_0 and ϵ_S are the dielectric constants of vacuum and solvent respectively. For a 0.1 M electrolyte solution, d is approx. 0.1 nm and the diffusion layer less than 1 nm. The charge access on the solvation side (Q) can then be estimated by equation 2.13.[38, p. 110]

CHAPTER 2. EXPERIMENTAL TECHNIQUES

$$Q = CE \tag{2.13}$$

Where E is here the galvanic potential difference between the metal and the interface.

Inside the electric double layer very high electric fields may be sustained. For concentrated electrolyte solutions the thickness of the double layer will be only 0.1 nm. For a galvanic potential difference of 0.1 V this corresponds to a field strength of about 10^9 V m^{-1}, which is strong enough to weaken chemical-molecular bonds. This phenomena is also known as *field dissociation*.

The Helmholtz model does not account for diffusion and was improved by the Stern model where electrostatic interactions are in competition with Brownian motion. For more information about the electrical double layer, the complete theory and its implications see ref. nr. [39, p. 544-556] and [40, p. 113-124].

2.1.3 Cyclic voltammetry experiments

Cyclic voltammetry is a useful technique to determine the redox properties of the soluble compounds such as 1-1 to 2-3. It allows indirect studies about the catalytic activity for CO_2 reduction to CO and NADH regeneration. In the presented studies a one-compartment cell was used for cyclic voltammetry experiments, either with a Pt or glassy carbon working electrode, a Pt counter electrode and a Ag/AgCl quasi reference electrode calibrated with ferrocene/ferrocenium (Fc/Fc$^+$) as an internal reference in organic solvents. In water based systems a standard calomel real reference electrode was used. For CO_2 reduction experiments, the cyclic voltammogram of the metal complexes under N_2 saturation was compared to the spectrum in the presence of carbon dioxide. Electrochemical experiments with organic solvents (e.g. acetonitrile) for CO_2 reduction were performed with 0.1 M tetrabutylammonium hexafluorophosphate (TBAPF$_6$) as supporting electrolyte, and in water based systems with a 0.1M KCl as supporting electrolyte. It was found that purging of the system with N_2 or CO_2 respectively, for about 15 min under stirring, is sufficient to achieve gas saturation of the electrolyte solution. The catalyst concentration in a cyclic voltammogram experiments was typically 1 mM and the CO_2 concentration was assumed to be at gas saturation of 0.28 M in acetonitrile.[41]

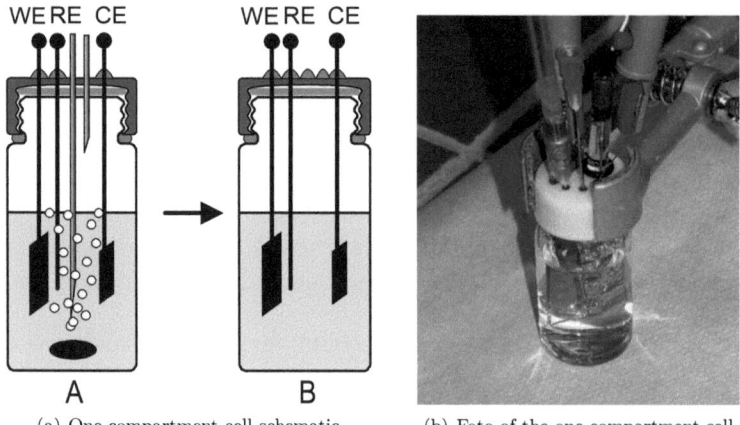

(a) One-compartment cell schematic   (b) Foto of the one-compartment cell

Figure 2.5: (a) One-compartment cell for cyclic voltammetry experiments during CO_2 purging (A) and during cyclic voltammetry measurement (B). Cells containing working electrode (WE), reference electrode (RE) and counter electrode (CE). (b) Foto of a typical one compartment cell with a three electrode setup contacted by crocodile clamps, Ag/AgCl real reference electrode in a 3 M KCl solution and in- and outlets for gas purging.

Scheme 2.5(a) shows a schematic picture of the one-compartment cell for cyclic voltammetry experiments during CO_2 purging (A) and during the voltammetry experiment (B). The one-compartment cell used for electrochemistry experiments typically contained about 14 ml of electrolyte solution and 10 ml gas phase. Cyclic voltammetry experiments were usually performed using a JAISSLE Potentiostat-Galvanostat IMP 88 PC.

Optical cyclic volltammetry experiments for photo-electrochemistry with organic semiconductors as working electrode were preformed in a quartz cuvette cell. A schematic representation of such a quartz cuvette for cyclic voltammetry experiments with organic semi conducting electrodes is depicted in scheme 2.6. These cells contain a transparent working electrode (WE), reference electrode (RE) and a counter electrode (CE). As light source a 35 W halogen spot lamp was used with white light irradiation. The distance between the cuvette and the light source was 7 cm if not stated otherwise.

Experiments with a quasi reference electrode (QRE) have to be calibrated

CHAPTER 2. EXPERIMENTAL TECHNIQUES

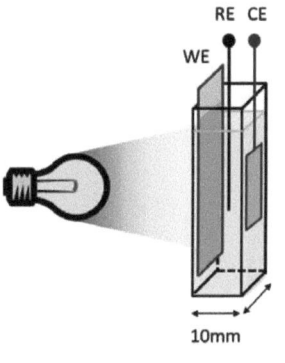

(a) Quartz cuvette schematic (b) Foto of the quartz cuvette setup

Figure 2.6: (a)Quartz cuvette for cyclic voltammetry experiments with organic semi conducting electrodes. Cells containing working electrode (WE), reference electrode (RE) and counter electrode (CE). (b) Foto of the Quartz cuvette electochemical setup inside the glove box under white light illumination of a 35 W spot lamp.

with ferrocene / ferrocenium (Fc/Fc^+) as an internal reference. The half-wave potential $E_{1/2\ Fc/Fc+}$ for Fc/Fc^+ was measured at eg. 375 mVvs.Ag/AgCl quasi reference electrode potential for the experiment shown in Figure 2.7. The recorded current-potential curve (black line with squares) is then recalculated according to the $E_{1/2\ Fc/Fc+}$ to the corresponding normal hydrogen electrode (NHE) potential (red line with circles). For the recalculation to NHE potential the $E_{1/2Fc/Fc+}$ vs. NHE potential offset was taken at 640 mV as suggested by Bazan et al. [42].

2.1.4 Controlled potential electrolysis

Controlled potential electrolysis experiments were generally performed in a gas tight one-compartment or H-cell with a Pt working electrode, a Pt counter electrode and a Ag/AgCl quasi reference electrode. Figure 2.8 shows a scheme of a typical H-cell setup used in the controlled potential experiments. Although a one-compartment cell is easier to handle in terms of setup building and sealing it has the big disadvantage that reduction products formed at the working electrode might be re-oxidised on the counter electrode. This problem can be avoided using an H-cell with separated anode and cathode part by a glass

2.1. ELECTROCHEMISTRY

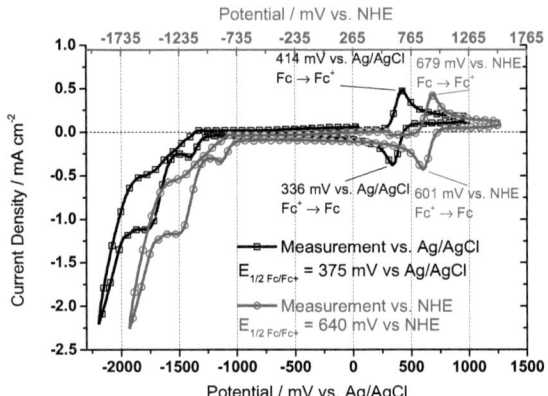

Figure 2.7: Typical example of a cyclic Voltammograms of 1-1 in CO_2 saturated electrolyte solution as measured during the experiment versus the Ag/AgCl quasi reference electrode potential (black lines with squares and recalculated according to $E_{1/2\ Fc/Fc+}$ to the NHE potential (red line with circles). Measurements are taken at a scan rate of $100\ \text{mVs}^{-1}$ in acetonitrile with TBAPF$_6$ (0.1 M), Pt working electrode, Pt counter electrode, and a catalyst concentration of 1 mM.

frit. However, since the expected product with the transition metal type of catalysts, in aprotic solvents like acetonitrile, is CO and the solubility of CO in a given solvent is typically in the order off 100 times less than for CO_2, it is expected that little to no CO will remain in solution. In fact the experiments performed with CO evolving catalysts did not show any difference concerning CO yield between the one compartment cell electrolysis experiment and the H-cell experiments.

Figure 2.9 shows a typical current density vs. time plot for CO_2-electrolysis experiment with compound 1-3 as performed in a one compartment cell. It can be seen that the reductive current initially dropped significantly and afterwards stayed constant with a small decay reflecting the declining concentration of CO_2 substrate in solution over the electrolysis time. Additionally, a plot of current density vs. $1/\text{time}^{1/2}$, as shown in Figure 2.9(b), can give useful information on the kinetics of the process. For example Fick's second law of diffusion shows that a linearity in the current vs. $1/\text{time}^{1/2}$ plot suggests both, a fast electron transfer rate and a time-independent surface concentration of the reactant (in this case CO_2) within the electrolysis time, compare Equation 2.14.[39]

CHAPTER 2. EXPERIMENTAL TECHNIQUES

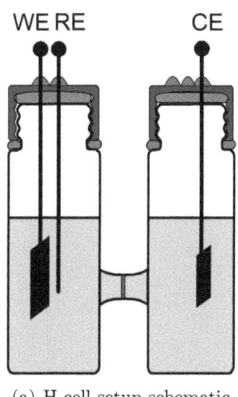
(a) H-cell setup schematic

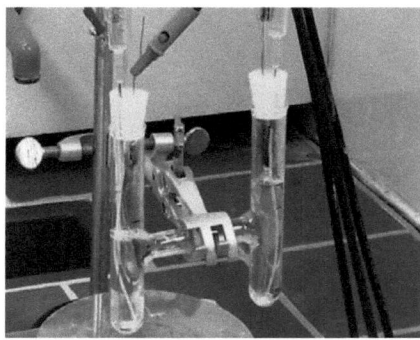
(b) Foto of H-cell

Figure 2.8: (a) H-cell setup with separated anode and cathode compartment for controlled potential electrolysis. Cells containing working electrode (WE), reference electrode (RE) and counter electrode (CE). (b) Foto of a typical H-cell used for CO_2 electrolysis experiments.

$$j = nF \left(\frac{D}{\pi}\right)^{1/2} \frac{c^0 - c^s}{t^{1/2}} \qquad (2.14)$$

Where j is the current density, n is the number of electrons, F is the Faraday constant, D is the diffusion coefficient, c^0 is the bulk concentration, c^s is the electrode surface concentration and t is the measurement time.

The slope of the line also allows calculating the diffusion coefficient D if the bulk and electrode surface concentrations are known. The equation cannot be applied for very short time scales, since the initial concentration gradient is very high and the current as such is not limited by diffusion, but by the electron transfer process, compare Figure 2.9(a).[38]

As a direct proof of the catalytic CO_2-reduction capability of the rhenium catalysts, headspace gas samples are taken and analysed with regard to the CO-concentration using GC and FTIR measurements. The application of transmission IR gas measurements, compared to standard GC analysis, has several advantages. IR measurement has a very short measurement time, high reproducibility, works at ambient temperatures and pressures, shows no vulnerability to interact with a mobile or stationary phase and the gas sample will not come in contact with the detector system. Details to this method used for

2.1. ELECTROCHEMISTRY

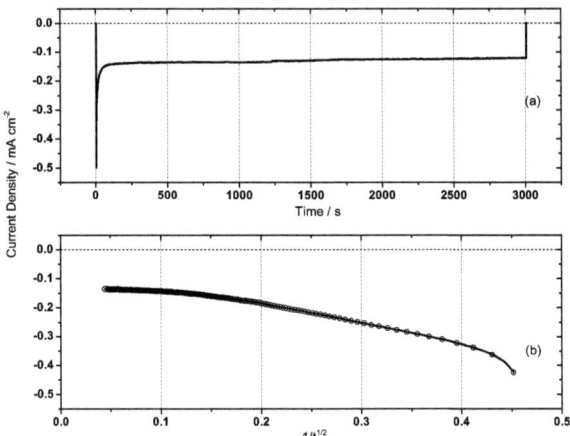

Figure 2.9: Typical current density vs. time plot for potentionstatic CO_2-electrolysis experiment of Re(5,5'-bisphenylethynyl-2,2'bipyridyl)$(CO)_3$Cl (1-3) at constant $-1950\,\mathrm{mV}$ vs.NHE, performed in acetonitrile solution saturated with CO_2 and an electrolysis time of 3000 s (a) and the same data plotted as current density vs. $1/\mathrm{time}^{1/2}$ (b).

CO measurements are described in the subsection *Optical spectroscopy methods* and have also been reported in a published paper.[43]

Knowing the amount of products formed and the charges passed through the electrochemical system, the Faradaic efficiency (η_F) can then be calculated according to Equation 2.15.

$$\eta_F = \frac{n_e \times n_P}{n_{etot}} \qquad (2.15)$$

Where n_P is the amount of products formed, n_e is the number of electrons needed for each product (compare reactions 1.1 to 1.8) and n_{etot} is the total number of electrons put into the system during the electrolysis experiment, usually optained by integration of the current time curve as plotted in Figure 2.9(a). For the Faradaic efficiency of CO formation Equation 2.15 can be rewritten into Equation 2.16

CHAPTER 2. EXPERIMENTAL TECHNIQUES

$$\eta_{FCo} = \frac{2 \times (n_{COgas} + n_{COsol})}{n_{etot}} \qquad (2.16)$$

where n_{COgas} is the amount of CO molecules formed in the gas phase, n_{COsol} is the amount of CO molecules in the electrolyte solution and 2 denotes the two electrons required for the CO_2 reduction to CO (according to the reaction1.2).

The number of molecules of CO in the gas phase is usually obtained by GC and FTIR analysis, while the number of molecules of CO dissolved in the electrolyte solution is more difficult to obtain. In the experiments presented herein the number was estimated using Henry's Law following Equation 2.17.

$$p_i = k_H \times c \qquad (2.17)$$

The Henry constant k_H is taken with (2507 atm mol$_{\text{solvent}}$ mol$_{CO}^{-1}$), derived from data of Castillo et al.,[44] p_i is the partial pressure of the solute CO and c is the concentration of CO in solution. The number of electrons consumed in the CO_2 electrolysis was determined by integration of the current-time curve of the electrolysis experiment.

With this approach, typical Faradaic efficiency for transition metal catalysts for the reduction of CO_2 to CO, by e.g. the compound 1-3, of about 43 % was deduced and ranks equally with the efficiencies that have been measured for compound 1-1.[43] Literature values of reported Faradaic efficiencies of similar or modified compounds reach values up to 100 %.[45, 46, 47] Furthermore, it has been shown that under these conditions (ACN:TBAPF$_6$) also small amounts of formate and oxalate can be formed, however with typical Faradaic efficiencies below 1 %.[45]

It should be stated that control experiments, e.g. without catalyst materials present under otherwise identical conditions, are important to verify the technique and should not yield detectable amounts of product formation.

2.1.5 Electropolymerisation

Electropolymerisation of the rhenium catalyst film in chapter 8 was done in potentiodynamic mode in a one-compartment cell also used for cyclic voltammetry experiments, either with a Pt or glassy carbon working electrode, a Pt counter electrode and a Ag/AgCl quasi reference electrode. For the formation of the rhenium catalyst film the catalyst monomer 1-3 was polymerized by repeatedly scanning to negative -1600 mV vs. NHE on a Pt working electrode at $50\,\text{mVs}^{-1}$ in nitrogen-saturated acetonitrile solution containing $0.1\,\text{M}$ $TBAPF_6$ and a monomer catalyst concentration of $2\,\text{mM}$.

Electropolymerisation of the polypyrrolle film in chapter 8 including the catalyst monomer 1-1 solution was done by potentiodynamic film formation of a pure polypyrrole film without compound 1-1 present. Pyrrole was electropolymerized on a Pt foil serving as supporting working electrode for the polypyrrole film. Pyrrole was used as received. $625\,\mu l$ of pyrrole were added to 18 ml of acetonitrile with $TBAPF_6$ ($0.1\,\text{M}$) as supporting electrolyte to receive a monomer concentration of $0.5\,\text{M}$. The electropolymerisation was performed by sweeping the potential between 1053 mV and -647 mV vs. NHE over 70 cycles with a scan rate of $100\,\text{mVs}^{-1}$.

In a subsequent similar experiment compound 1-1 was dissolved in the pyrrole monomer electrolyte solution with a concentration of 2 mM and used for electro polymerization. After electropoylmerization at constant current of 0.1 mA the formed film was removed from the system and washed with pure acetonitrile solution to remove any initial monomer and catalyst material not incorporated into the polymer matrix. Then the polymer film was used as working electrode for heterogeneous CO_2 reduction.

The potentiodynamic electropolymerization of thiophene compounds in chapter 9 was performed in different solvents. 0.1 M thiophene to polythiophene (8-1) was polymerized in ACN solution containing 0.1 M $TBAPF_6$ as supporting electrolyte. The cyclic voltammograms were recorded at a scan rate of $100\,\text{mVs}^{-1}$ inside the glove box using a Pt working electrode (WE), a Pt counter electrode (CE) and a Ag/AgCl quasi reference electrode. Potentiodynamic electropolymerization of 0.056 M 3-hexylthiophene to poly (3-hexylthiophene) (8-2) was performed in propylene carbonate solution containing 0.1 M $TBAPF_6$ as

CHAPTER 2. EXPERIMENTAL TECHNIQUES

supporting electrolyte. The potentiodynamic electropolymerization of approx. 0.015 M Re(4-methyl-4'-(7-thienylheptyl)-2,2'-bypyridene(CO)$_3$Cl to fac-(2,2'-bipyridyl)Re(CO)$_3$Cl functionalized poly (3-hexylthiophene) (8-3) was done in in BFEE solution.

2.2 Spectroscopic methods

2.2.1 UV-Vis absorption and PL spectroscopy

UV-vis absorption measurements for the monomer catalysts were performed in a 1 cm quartz glass cuvettes at 298 K with a Cary 3G UV-visible spectrophotometer. Light absorption of the rhenium catalyst film on a Pt working electrode was characterized using an Ocean Optics fibre spectrometer and an ISP-R integrating sphere. The difference in the diffuse reflectance spectra of the non-transparent electrode was compared to a white light standard.

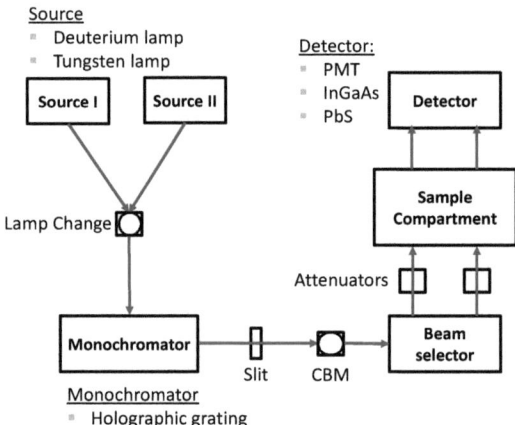

Figure 2.10: Illustration of the general UV-Vis spectrometer setup consisting of a deuterium (D2) and tungsten light source, monochromator, the beam selector, sample compartment and the detector.

Transmission measurements for the PL quenching experiments described in chapter 9 were taken on a Perkin Elmer LAMBDA 1050 double monochromator spectrometer (source doubling mirror) between 400-700 nm in 2 nm steps, with a slit width of 2 nm and a detector response time of 0.2 second. When required, signal to noise was optimized by attenuating the reference beam with internal attenuators. A schematic illustration of the spectrometer is depicted in Figure 2.10.

The LAMBDA 1050 uses automatic 2 A and 3 A attenuation and can

CHAPTER 2. EXPERIMENTAL TECHNIQUES

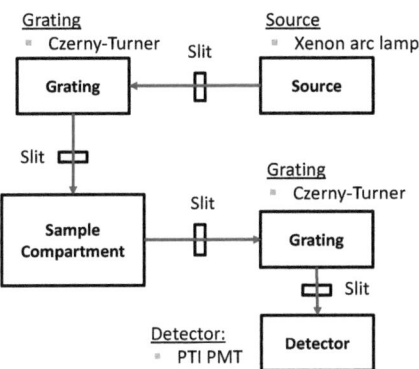

Figure 2.11: Illustration of the general PL spectrometer setup consisting of a Xenon arc source, Czerny-Turner type gratings, sample compartment and the PTI PMT detector.

be similarly amplified. For all spectra, autozero (100 % and 0 %) correction scans were taken (baseline correction). The spectrometer is equipped with a deuterium (D2) lamp as the UV and a tungsten lamp as the visible and near infrared (NIR) source. The energy is dispersed into specific wavelengths by the use of a reflective gratin. The source change is set to 320 nm, the common beam mask (CBM) is set to 100 %.

The LAMBDA 1050 utilizes a Photomultiplier R6872 and three detectors for energy detection from 175-3300 nm. A gridless PMT for detection in the UV/Vis, a Peltier cooled InGaAs detector for use in the 800-2600 nm region and a Peltier cooled PbS detector for the range from 2500-3300 nm.

Photoluminescence (PL) spectra were recorded on a PTI QuantaMaster 400 Spectrofluorometer with a continuous Xenon arc lamp (75 W) light source emission range from 185 nm to 680 nm, a Czerny-Turner type excitation monochromator (throughput 65 % at 300 nm) with a focal length of 300 mm, a excitation grating with 1200 line/mm (300 nm blaze), emission grating with 200 line/mm (400 nm blaze) and a multi-mode PTI PMT detector model 914, with a spectral response from 185 to 900 nm (Quantum Efficiency at 260 nm (Peak) 25.4 % typ.). A schematic illustration of the spectrometer is depicted

2.2. SPECTROSCOPIC METHODS

in Figure 2.11.

2.2.2 FTIR absorption spectroscopy

FTIR measurements were performed on a BRUKER IFS 66/S FTIR spectrometer with $4\,\text{cm}^{-1}$ spectral resolution. Scheme 2.12 shows a very general illustration of the FTIR setup consisting of a mid infrared (MIR) Glowbar light source, an interferometer with a KBr beam splitter, the sample chamber and the mercury cadmium telluride (MCT) and triglycine sulfate (DTGS) detector.

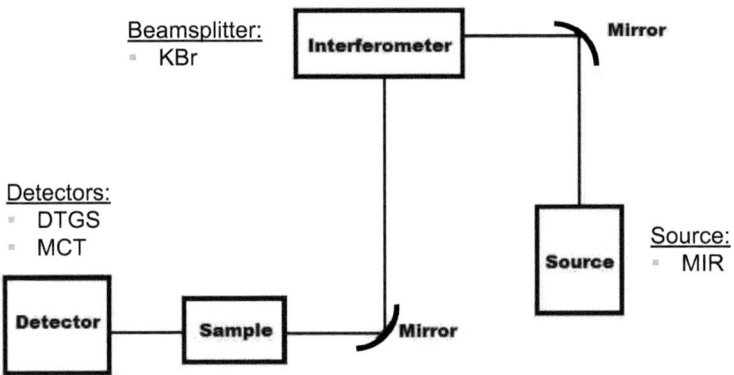

Figure 2.12: Illustration of the general FTIR spectrometer setup consisting of a MIR light source, an interferometer, the sample chamber and the detector.

For Faradaic efficiency calculations, FTIR gas analysis was used as a complementary and independent technique to standard GC measurements for identifying the reduction products formed. For the FTIR measurements, a gas tight transmission cell (see Scheme 2.13) with ZnSe windows was designed to measure IR absorption in transmission mode. The spectra were recorded with and without 5 ml headspace sample after different times of electrolysis of the catalyst compounds in solution. For calibration of the systems, a defined amount of calibration gas (Linde) containing 11 % CO_2, 1.5 % CO and 600 ppm C_3H_8 in N_2 was used.

This method has, compared to standard GC analysis, several advantages. The IR measurement has a very short measurement time, high reproducibility, works at ambient temperatures and pressures, shows no vulnerability to interact with a mobile or stationary phase and the gas sample will not come in contact with the detector system.

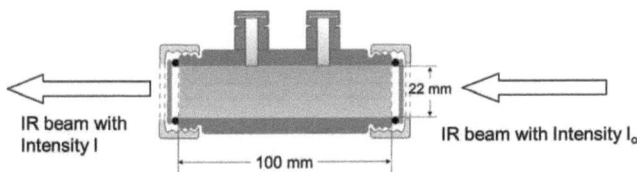

Figure 2.13: Illustration of the the gas tight transmission cell with ZnSe windows built for IR difference absorption spectrum measurement in transmission mode, where I_0 denotes the intensity of the incomeing IR beam and I the intensity of the IR radiation after the sample.

Figure 2.14 shows a typical IR difference absorption spectrum in transmission mode. The two peaks centred around $2143\,\text{cm}^{-1}$ correspond to the infrared active rotational-vibrations of the P and R branch of gaseous CO. For further details on the infrared active vibrations of various molecules see reference [48]. The absorption increase at $2350\,\text{cm}^{-1}$ and the two double peaks centred around $3650\,\text{cm}^{-1}$ correspond to the infrared active vibrations of CO_2.

The area under the peak can be determined by integration. Figure 2.15 shows the calibration of the system by difference absorption spectra in transmission mode, with different volumes of a calibration gas mixture (11 % CO_2, 1.5 % CO and 600 ppm C_3H_8 in N_2). It was found that the peak area is linearly depending on the concentration of CO in the system (see Figure 2.15). With this setup, it was possible to determine the CO gas concentration in an initially N_2-filled transmission cell as low as 500 ppm.

2.2. SPECTROSCOPIC METHODS

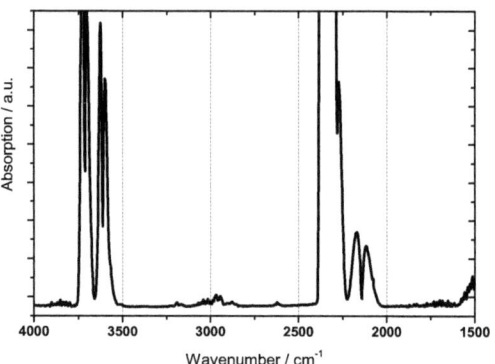

(a) IR difference absorption spectra

(b) Foto of IR transmission gas cell

Figure 2.14: (a) IR difference absorption spectrum in transmission mode, with and without 5 ml headspace sample after 50 minutes electrolysis experiment of compound 1-3 in solution. (b) Foto of the IR transmission gas cell with two ZnSe windows as it is mounted in the IR spectrometer compartment.

CHAPTER 2. EXPERIMENTAL TECHNIQUES

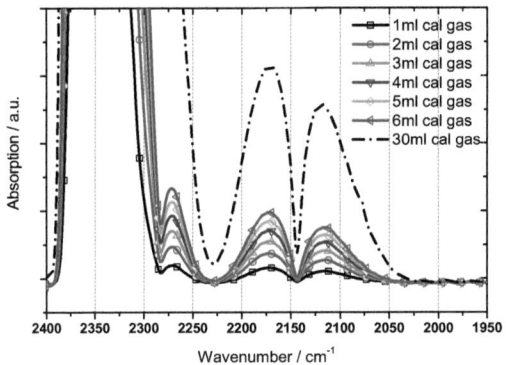

(a) IR difference absorption spectra

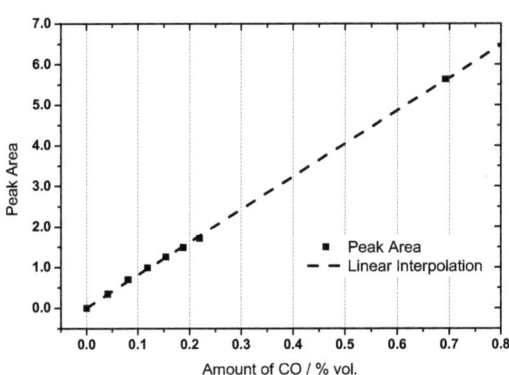

(b) Peak area of the spectra in (a)

Figure 2.15: (a) IR difference absorption spectra in transmission mode, with different volumes of a N_2 calibration gas mixture (11 % CO_2, 1.5 % CO and 600 ppm C_3H_8 in N_2). (The two peaks centred around $2143\,\text{cm}^{-1}$ correspond to the infrared active vibration of CO, the peak at $2350\,\text{cm}^{-1}$ to the vibrations of CO_2.). (b) Peak area of IR difference absorption peaks centred around $2143\,\text{cm}^{-1}$ in transmission mode at different CO volumes (black squares) and linear interpolation between peak areas (black dotted line).

2.2.3 ATR-FTIR absorption spectroscopy

Infrared measurements in attenuated total reflection (ATR) mode were performed on a BRUKER IFS 66/S FTIR spectrometer at room temperature, using a mercury-cadmium telluride (MCT) detector cooled with liquid nitrogen prior the measurements. For all ATR-FTIR measurements ZnSe crystal was used as reflection element, which was pre-cleaned by polishing with diamond paste (1 and 0.25 µm) and additional rinsing in a reflux system with acetone. For electropolymerisation, a thin (10 nm) layer of Pt was sputtered onto the ZnSe crystal, serving as a transparent working electrode (WE).

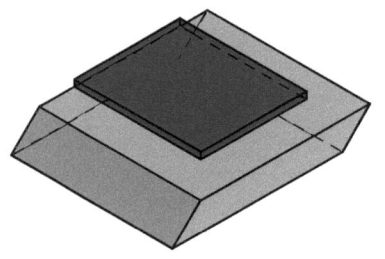 

(a) Schematic of an ATR-FTIR ZnSe crystal   (b) Foto of an ATR-FTIR ZnSe crystal

Figure 2.16: (a) Schematic of a ATR-FTIR ZnSe crystal as reflection element. For spectroelectrochemistry measurements, a thin (10 nm) layer of Pt was sputtered onto the ZnSe crystal, which served as a transparent WE. (b) Foto of an ATR-FTIR ZnSe crystal with a thin (10 nm) layer of sputtered Pt and conductive silver paste on top for back contact

Scheme 2.16 shows the schematic of a ATR-FTIR ZnSe crystal as reflection element. Since ZnSe is not conducting, for spectroelectrochemistry measurements, a thin (10 nm) layer of Pt was sputtered onto the ZnSe crystal, which served as a transparent WE.

For the electropolymerization of compound 1-3 the polymer growth process was investigated with an ex-situ ATR-FTIR technique. For this measurement, the ZnSe reflection element covered with a thin (10 nm) sputtered film of platinum was used as working electrode, and a 150 nm layer of the rhenium catalyst film was potentiostatically electropolymerized on the Pt-surface of the modified ZnSe ATR crystal. The experiment was performed in a one compartment electrochemical cell as depicted in Scheme 2.17(A).

CHAPTER 2. EXPERIMENTAL TECHNIQUES

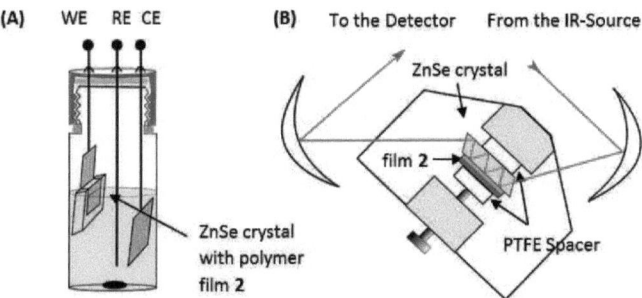

Figure 2.17: A) One-compartment cell with a ZnSe reflection element covered with a thin (10 nm) sputtered layer of Pt (gray) as the WE for the electropolymerization of the catalyst polymer (2, violet). B) Schematic of the mounted ATR-FTIR crystal with the polymer film (violet) inside the spectrometer.

The electrolyte solution contained 0.1 M TBAPF$_6$, initial monomer 1-3 concentration of 2 mM in acetonitrile. For the film formation, the electrochemical cell was connected to the potentiostat and a constant potential of −1550 mV vs. NHE was applied for 500 seconds. After electropolymerization, the ZnSe/Pt electrode with the 150 nm thick catalyst film was washed with pure acetonitrile and dried under N$_2$ atmosphere. Then the crystal was mounted into an ATR-FTIR setup between two PTFE spacers (as depicted in Scheme 2.17(B)) and the ATR-FTIR difference absorption spectra between a pure ZnSe/Pt electrode and the ZnSe/Pt electrode with the catalyst film were recorded. (The corresponding spectra are shown and discussed in chapter 8.)

2.3 Analytical methods for product analysis

For a direct proof of the catalytic CO$_2$ reduction capability of the catalyst materials, product analytics is a very important point. In most cases of the CO forming catalytic compounds 1-1 to 2-3, headspace gas samples were taken and analysed regarding the CO concentration by either GC or FTIR headspace gas sampling. For the FTIR absorption measurements a special gas tight IR transparent measurement cell containing two ZnSe windows was developed in our laboratory as already described (for a schematic drawing see Scheme 2.13). Additionally to the FTIR measurements gas chromatography was the main

2.3. ANALYTICAL METHODS FOR PRODUCT ANALYSIS

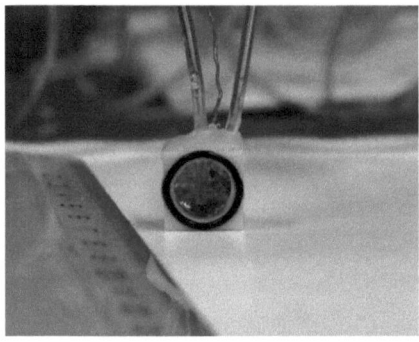 

(a) One-compartment cell (b) ATR-FTIR setup

Figure 2.18: (a) Picture of the PTFE cell for ATR-FTIR spectroelectrochemistry with tubes for electrolyte in- and outlet and the Pt counter electrode visible in the back of the cell (b) Foto of the ATR-FTIR setup mounted in the FTIR spectrometer with the three electrode setup and the ZnSe ATR crystal.

technique for product determination.

2.3.1 Gas chromatography

Gas chromatography (GC) analysis for CO was conducted on a Thermo Trace GC equipped with a TCD detector and a Phenomenex PLTT 5A column (30 m, 0.53 mm ID, 25 μm film). The carrier gas was helium at 3 ml min^{-1}, the GC was programmed from 45 °C for 5 min to 300 °C for 1 min with a heating rate of 30 °C min^{-1}. The injector was operated at 300 °C with a split ratio of 1:10 and the detector at 200 °C with 27 ml min^{-1} make up gas. 1 mL sample was injected with a gastight syringe directly from the reaction vessel.

Figure 2.19(a) shows a typical GC analysis of a headspace sample (200 μl) after a 25 minutes electrolysis experiment with compound 1-3 in a single-compartment cell containing CO_2 saturated acetonitrile solution with TBAPF$_6$ (0.1 M), a Pt working electrode, a Pt counter electrode, and a catalyst concentration of 1 mM. The chromatogram is shown in the time frame of 0 to 550 s. In this measurement three peaks are present corresponding to O_2 and N_2 at around 60 s, CO at around 90 s and CO_2 at 500 s retention time. Figure 2.19(b) shows liquid GC calibration measurements for various concentrations of

CHAPTER 2. EXPERIMENTAL TECHNIQUES

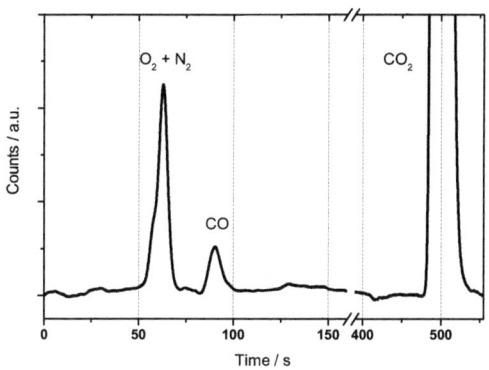

(a) Gas Chromatography

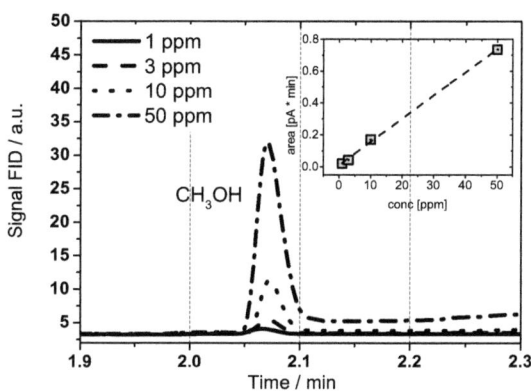

(b) Liquid Gas Chromatography

Figure 2.19: (a) GC analysis of a headspace sample (200 µl) after a 25 minutes electrolysis experiment with compound 1-3 in a single-compartment cell containing CO_2 saturated acetonitrile solution with $TBAPF_6$ (0.1 M), Pt working electrode, Pt counter electrode, and a catalyst concentration of 1 mM. (b) Liquid GC calibration measurements for various concentrations of methanol in water. The retention time for the methanol peak maximum is typically at 2.07 min. Inset: Linear fit of the calibration peak area.

methanol in water with an injection volume of 1 µl. The retention time for the methanol peak maximum is typically at 2.07 min. The inset shows the linear polynomial fit of the calibration peak area according to y = intercept + slope · x, with a slope of 0.015 with an error of 3.404^{-4} and the intercept at 0.01 with an error of 0.009. The calibration measurement shows excellent linearity in the concentration range from 1 to 50 ppm of methanol with a mean square value (R^2) of 0.99891 and a Pearson R of 0.99945.

2.3.2 Ion chromatography

Ion chromatography (IC) measurements for the detection of formate, acetate, carbonate and oxalate were done on a Dionix ICS 5000 equipped with a conductivity detector and a separated pre- (AG19, CAP, 0.4 x 50mm) and main-column (AS19, CAP, 0.4 x 250mm). The eluent source was a Dionix EGC-KOH (capillary). Measurement time was 27 min. with 45 mM target concentration and a reduced concentration of 10 mM (6 min). Eluent generation for 10min (10mM) with increasing the concentration (ramp 3 mM/min) for 15 min to 45 mM (3 min). The sample (1 ml) was injected with a 2 or 5 ml plastic syringe (Braun) using a PES 0.45 µm Roth filter.

2.4 Homogeneous photocatalysis

Photocatalytic reactions were performed in a 10 ml quartz cell filled with 6 ml of a mixture of dimethylformamide and triethanolamine (DMF:TEOA = 5:1/v:v) containing a catalyst concentration of 2.6 mM under CO_2 saturation. The solution was stirred during the experiment and a 360 nm low pass cutoff filter was used. The irradiation time was 18 h under 26900 lux with an Osram 400 W Xenophot xenon lamp. The irradiated cell cross section area was 3.14 cm². Gas samples were taken with a gas tight syringe and transferred to eighter FTIR gas analysis (for a schematic drawing see Scheme 7.2, in chapter 7) or GC analysis.

CHAPTER 2. EXPERIMENTAL TECHNIQUES

2.5 Catalyst materials

Scheme 2.20 and 2.21 shows the schematics of four rhenium(I) tricarbonyl chloride complexes with different diimine ligand systems, that is (2,2'-bipy.)Re(CO)$_3$Cl (1-1), (4,4'-dicarboxyl-2,2'-bipy.)Re(CO)$_3$Cl (1-2), (5,5'-bisphen.-2,2'-bipy.)-Re(CO)$_3$Cl (1-3) and [5,5'-bis (LL)-2,2'-bipy.]Re(CO)$_3$Cl (1-4) where LL for compound 1-4 is (2,6-bis-octyloxy-4-formyl)phenylethinyl) used for CO_2 reduction throughout the thesis. Scheme 2.20(b) shows the schematics of three different Re(BIAN-R)(CO)$_3$Cl compounds (2-1 to 2-3) that have been used as novel materials for the aim of CO_2 reduction in chapter 6. The compounds have been synthesized according to literature procedure. [49]

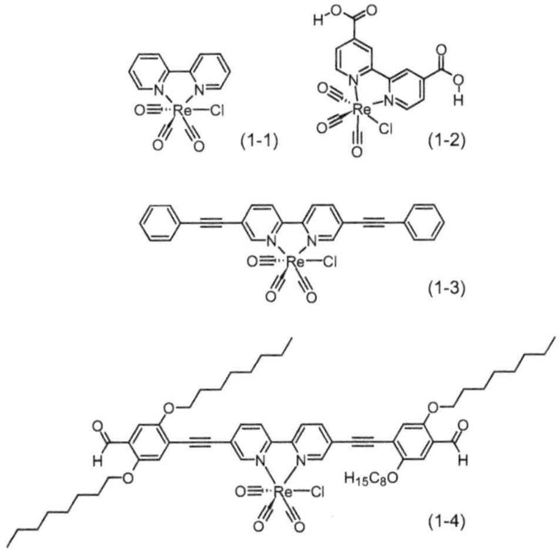

Figure 2.20: Schematic chemical structures of different rhenium bipyridyl compounds (2,2'-bipyridyl)Re(CO)$_3$Cl (1-1), (4,4'-dicarboxyl-2,2'-bipyridyl)Re(CO)$_3$Cl (1-2), (5,5'-bisphenylethynyl-2,2'-bipyridyl)Re(CO)$_3$Cl (1-3) and [5,5'-bis ((2,6-bis-octyloxy-4-formyl)phenylethinyl)-2,2'-bipyridyl]Re(CO)$_3$Cl (1-4) for CO_2 reduction.

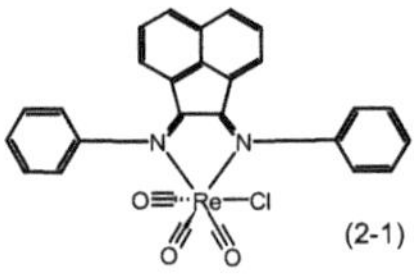

2.5. CATALYST MATERIALS

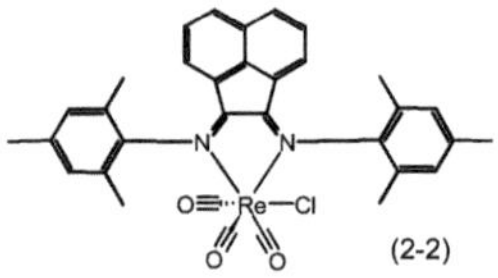

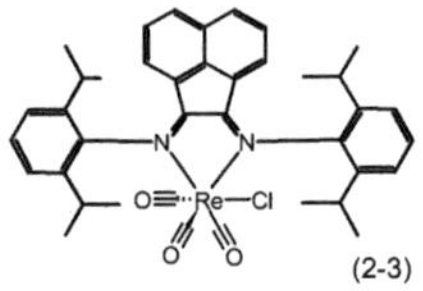

Figure 2.21: Schematic chemical structures of three different rhenium compounds with bis(arylimino)acenaphthene derivatives (BIAN-R) ligands (2-1), (2-2) and (2-3) for CO_2 reduction.

Chapter 3

On the nature of rhenium-(I) bipyridine complexes

3.1 Symmetry and structure

Octahedral complexes have a coordination number of 6, meaning that there are six places around the metal center where ligands can bind. Organometallic rhenium-(I) complexes form such an octahedral molecular geometry wherein six groups of atoms or ligands are symmetrically arranged around a central rhenium metal center (M) atom. In this configuration rhenium (Re^{+1}) has the electron configuration d^6 similar to Mn^{+1}, W^0 or Fe^{+2}.

In an octahedral molecular geometry (Octahedral point group of symmetry, O_h) the complex demonstrates several symmetry elements that can be summarized as:

- Six 4-fold rotation axes ($6C_4$), emerging from each pair of opposite apices.
- Eight 3-fold rotation axes ($8C_3$), emerging from each pair of opposite triangular faces.
- Six 2-fold rotation axes ($6C_2$), emerging from the center for each pair of edges.
- Three horizontal mirror planes ($3\sigma_h$), perpendicular to principal axis.
- Six vertical mirror planes ($6\sigma_d$), parallel to C_4 and bisecting two C_2' axes.

3.1. SYMMETRY AND STRUCTURE

- Center of symmetry (the inversion point is at the center of the octahedral).
- Additionally there are Improper axis (S_n) as six S_4 and eight S_6.

See Figure 3.1 for illustration.[50, p. 404-425]

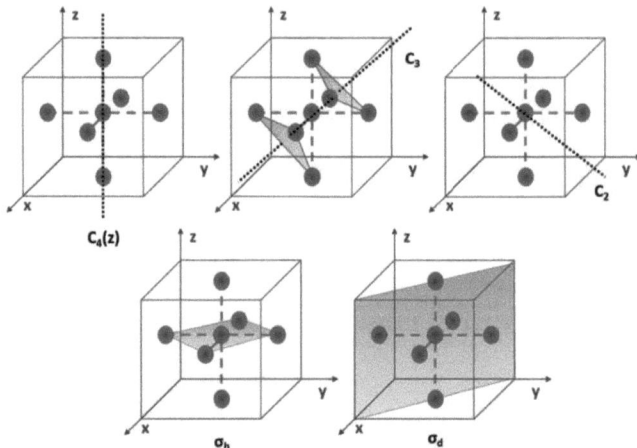

Figure 3.1: Exemplary illustration of symmetry elements C_4, C_3, C_2, σ_h and σ_d in an octahedral molecular geometry.

When two or more identical ligands are coordinated to an octahedral metal center (M), i.e. rhenium-(I)bipyridine(bpy) complexes, the complex can form isomers. Isomers are molecules that demonstrate the same composition and molecular formula, but differ in their structures which may lead to differences in their physical and chemical properties.[51, p. 607-617]

The naming system for these isomers depends upon the number and arrangement of different ligands and is illustrated in Figure 3.2.

For two ligands, the system can arrange in the *cis* and *trans* configuration as shown in Figure 3.2(a), where isomers are *cis*, if the ligands are commonly neighboring, and *trans*, if the ligand groups are situated 180° to each other. Different for three ligands, compare Figure 3.2(b), the system can arrange in a facial (*fac*) or meridional *mer* configuration. In this configuration the facial isomer (*fac*) consist of three identical ligands that occupy one face of the octahedron surrounding the metal atom, so that they form a face or plane. The

CHAPTER 3. ON THE NATURE OF RHENIUM-(I) BIPYRIDINE COMPLEXES

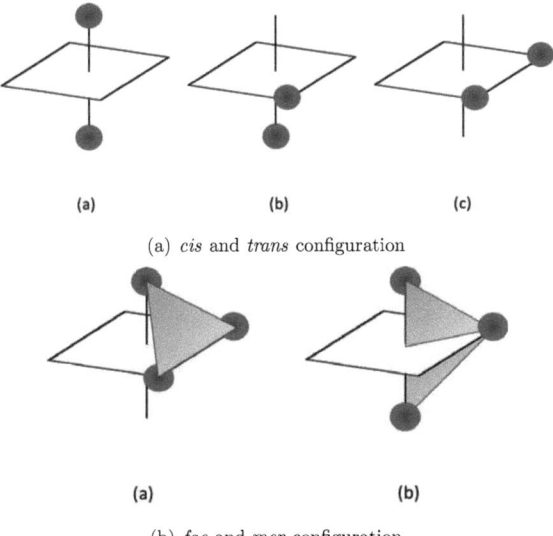

(a) *cis* and *trans* configuration

(b) *fac* and *mer* configuration

Figure 3.2: (a) Schematic representation of *cis*- and *trans* configuration for two identical ligands. a in trans, b and c in cis configuration. (b) Schematic representation of *fac* and *mer* configuration for three identical ligands. a in *fac* and b in *mer* configuration.

51

meridional isomer (*mer*) is described by a set of three identical ligands that occupy a plane passing through the metal atom. For four identical ligands the system can then again arrange in a *cis-trans* configuration.[51, p. 607-617]

3.2 The photostability of rhenium-(I) bipyridine complexes

If one is interested in the photostability of rhenium-(I) bipyridine (bpy) complexes one has to look at the interactions of the ligands of the complex with the d-electron orbitals of the central rhenium metal atom. The d-orbitals in a pure rhenium metal atom are usually degenerate in nature. Interactions between the electrons of the ligands and those of the metal center produce a crystal field splitting where the d_{z^2} and $d_{x^2-y^2}$ orbitals increase in energy, while the other three orbitals of d_{xz}, d_{xy}, and d_{yz} decrease in energy. If the rhenium metal (or any metal d-orbitals) is occupied in an octahedral ligand field, the five-fold degenerate metal d-orbitals split in two fold degenerate e_g and three fold degenerate t_{2g} orbitals. The so formed e_g-orbitals are higher in energy because their electron density forms a maximum along the octahedral direction.[52, p. 140]

The intensity of the ligand field splitting depends on the central metal atom (or ion) and the nature of the ligand. Strong ligands like CO lead to a significant splitting of the metal d^6 orbitals in rhenium (Re^{1+}) as illustrated in Figure 3.3 below.[51, p. 607-617]

CO is a ligand molecule with negative partial charge (δ^-) at the oxygen and a positive partial charge (δ^+) at the carbon atom. The bond is a triple bond formed via one sigma- and two pi-bonds. The bond length distance d_{CO} is about 113 pm and it is almost nonpolar with a dipole moment of about 0.11 Debye. The molecule shows typical infrared vibrations around 2143 cm^{-1}.[53] CO is a neutral ligand in the rhenium-(I)bipyridine(bpy)(CO)$_3$Cl complexes, different to Cl$^-$, which is a negatively charged ligand.

For the d^6-low-spin-configuration rhenium-(I) bipyridine (bpy) complexes, as illustrated in Figure 3.3, the ligand-field stabilization energy is very high, since all six electrons occupy the energetically lower t_{2g} orbitals. This is one reason for the primarily octahedral configuration of Re^{1+} complexes.[51, p. 607-

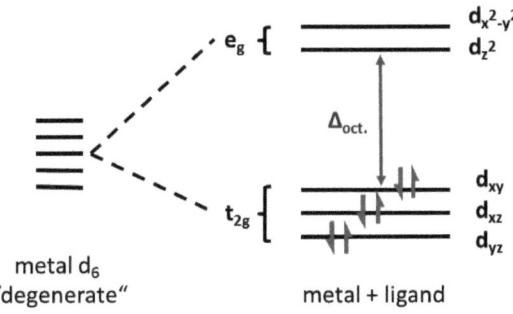

Figure 3.3: Schematic illustration of the splitting of the five-fold degenerate metal d^6-orbitals in two fold degenerate e_g and three fold degenerate t_{2g} orbitals within an octahedral ligand field.

617]

In solution, the d^6-orbital splitting in an octahedral ligand field is influenced by the solvent polarity. Rhenium complex demonstrate typically a negative solvatochromism, which means that the transition energy or wavelength between t_{2g} and e_g orbitals is shifted to shorter wavelength (λ) with increasing solvent polarity.[54, 55, 56]

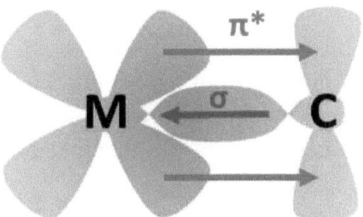

Figure 3.4: Schematic illustration of the CO molecule forming a sigma donor bond with the rhenium metal center d-orbitals and act as a π^*-acceptor, where the π-orbitals of the CO make a back-bonding (back donation) to the metal d-orbitals

When the Cl^- ligand approaches the rhenium metal center d-orbitals they form a sigma bond where the Cl^- ion serves as sigma donor. The CO molecule also forms a sigma-donor bond with the rhenium metal center d-orbitals, but also acts as a π^*-acceptor, where the π-orbitals of the CO make a back-bonding (back donation) to the metal d-orbitals, compare Figure 3.4 for illustration. Bipyridine (bipy) ligands have a similar back-bonding mechanism and also contribute two electrons pairs which makes them so called bidentate ligands. This backbonding increases the stability of the CO and bipy type of ligands over the Cl^- ligands significantly and results in a photostable complex where the Cl^- bonding is broken first upon excitation. This is important for photocatalysis, since it opens a free ligand side upon Cl^- release for the CO_2 to interact with the rhenium complex.

3.3 The Jablonski-Diagramm for rhenium-(I) bipyridine complexes

The antibonding d-metal orbitals would be unstable, however, due to the strong LF-splitting, the antibonding d-orbitals are energetically higher than the $\pi*$-bipy orbitals (compare also Jablonski diagram Figure 4.1 in the thesis). That is why the MLCT is the lowest lying excited state in rhenium-(I) bipyridine (bpy) complexes and is stable. While the metal centered (MC) transition from d (t_{2g}) to d (e_g) orbitals is typically in the UV-range, the MLCT transition is well in the visible region around a wavelength of 450 nm of the light spectra, compare Figure 3.5(a).

Figure 3.5(b) shows a schematic representation of a typical absorption spectra of rhenium-(I)bipyridine$(CO)_3Cl$ complexes. At high energies the metal centered (MC) d-d transitions are observed around 200 nm, with very low intensities of $\epsilon \approx 10\text{-}100$, where ϵ is the absorptivity of the species under investigation according to the Beer-Lambert law shown at equation 3.1.

$$T = \frac{I}{I_0} = e^{-\epsilon l c} \qquad (3.1)$$

In this equation, T is the transmittance, I_0 and I are the intensities of the incident and transmitted radiation, l is the distance the light travels through the material and c is the concentration of the species under investigation.[57]

As can be seen in Figure 3.5(b) the intensities of the metal centered (MC) d-d transitions are rather low because they are *parity forbidden* transitions. The parity forbidden transitions become partly allowed due to a mixing of the d-orbitals with other orbitals, such as p- or f-orbitals. At wavelengths around 300 nm the intraligand (IL) bipyridyl π-π^* transitions are observed with high intensities ($\epsilon\approx$100 000). Sometimes a week shoulder at longer wavelength around 350 nm is visible that can be attributed to a metal-to-ligand charge transfer (MLCT) from rhenium to the CO-ligands.[43] At about 450 nm the MLCT from rhenium to bipyridine is observed with intensities of $\epsilon\approx$10 000. The width of the UV-Vis absorption bands of liquid samples arises from their vibrational structure that can usually not be resolved in condensed phases.[53]

3.3. THE JABLONSKI-DIAGRAMM FOR RHENIUM-(I) BIPYRIDINE COMPLEXES

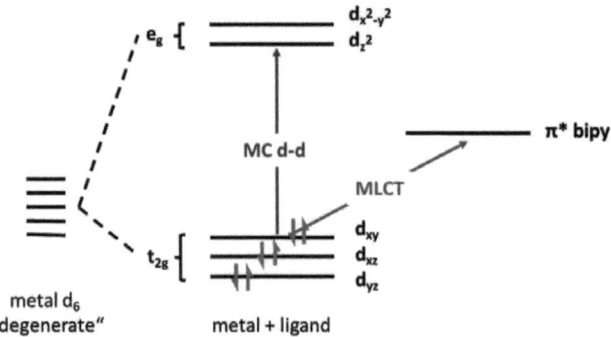

(a) MC transtions from $d(t_{2g})$ to $d(e_g)$ and MLCT transitions

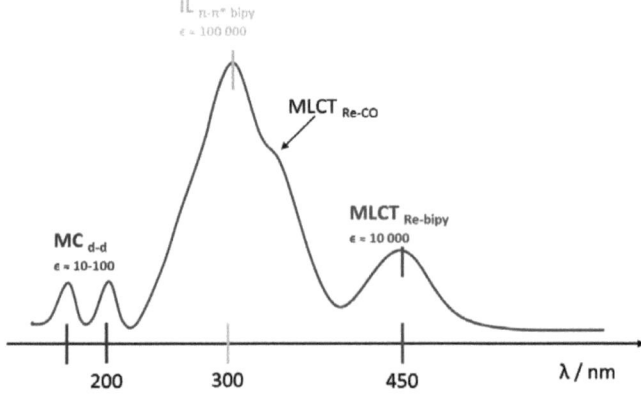

(b) Typical absorption spectra of rhenium-(I)bipyridine$(CO)_3$Cl

Figure 3.5: (a) Schematic illustration of the metal centered (MC) transtions from $d(t_{2g})$ to $d(e_g)$ and the metal to ligand charge transfer (MLCT) from the metal d-orbitals to the π^*-orbitals of the bipyridine ligand. (b) Schematic representation of a typical absorption spectra of rhenium-(I)bipyridine$(CO)_3$Cl complexes.

Chapter 4

Photophysical results

The photocatalytic efficiency of different diimino rhenium carbonyl complexes for the reduction of carbon dioxide to CO is strongly dependent on the electronic structure and the redox properties of these compounds. As stated by Takeda et al.[58] the lowest electronic excited states of bipyridine-based rhenium diimime carbonyl complexes, mainly of ^{3}MLCT* and $^3\pi\pi^*$ character, are frequently quite close in energy and may also be mixed with each other. Focusing on photochemical CO_2 reduction, almost all applicable catalysts have in common, that their lowest excited state is of the ^{3}MLCT* character with a relatively long lifetime in the 10-100 ns range.[58] This state can be quenched reductively by suitable sacrificial electron donors such as triethanolamine (TEOA or TEA) to generate a one-electron-reduced (OER) species. These OER molecules can form adducts with CO_2, the structure of which is not yet fully clear. There are various proposed intermediates by different research groups.[59, 60, 61] In most of the suggested catalytic pathways, one of the OER-CO_2 adducts reacts with a second OER radical in order to obtain CO as a reaction product and retrieve the initial catalyst.

The nature of the bipyridyl ligand influences the energetic levels of the possible electronic transitions upon excitation with UV and visible light. A simplified scheme of the most common electronic structure for similar rhenium carbonyl complexes is shown in Scheme 4.1.[58] The prototype compound Re(bpy)(CO)$_3$Cl (1-1) is taken as a basic template for the following descriptions. It's UV-visible absorbance spectrum in toluene is shown in Figure 4.3 (black line with squares). It shows two distinct maxima at 300 and 400 nm

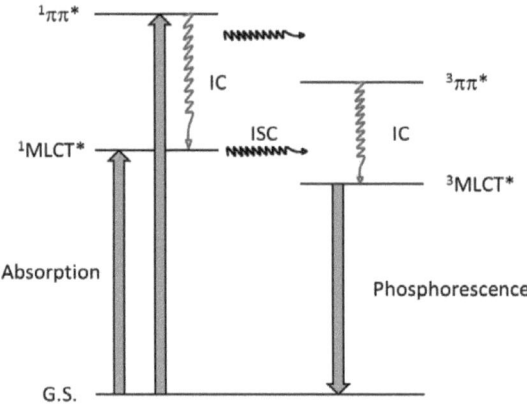

Figure 4.1: Schematic energy diagram of the lowest-lying excited states of complexes 1-1 to 1-4.[58]

and the room temperature photoluminescence (red line with triangles) at an excitation wavelength of $\lambda_{exc} = 380$ nm shows a broad emission from 450 to 750 nm with a maximum at 550 nm.

The absorbance spectra of the rhenium bipydridyl complexes 1-1 to 1-4 exhibit strong electronic $^1\pi\pi^*$ intraligand transitions of the diimine ligands in the higher energetic region, usually at wavelengths shorter than 330 nm and MLCT signatures at lower energies.[62, 63, 49, 64] Additional weaker UV-bands of IL origin can be observed for compounds 1-3 and 1-4 in the 300 - 320 nm spectral region. These bands suggest energetic splitting of approximately 2200 cm^{-1} due to coupling to phenylethynyl C-C vibrational modes.[62] The absorbance and emission data in various solvents for compounds 1-1 to 1-4 are summarized in Table 6.1.

Besides substituent effects on the spectra, what is most remarkable and also well known about the absorbance of the described compounds is the fact that the ^{1}MLCT transition is strongly dependent on solvent effects;[54, 62, 65] this can also be seen in Figure 4.2 comparing the spectra of compound 1-3 in dichloromethane to the spectrum recorded in acetonitrile. In the more polar solvents such as methanol or acetonitrile the lowest lying ^{1}MLCT absorbance is shifted to the red by 30 nm compared to less polar solvents like toluene or dichloromethane. Furthermore the carboxyl substituted compound 1-2 shows a

CHAPTER 4. PHOTOPHYSICAL RESULTS

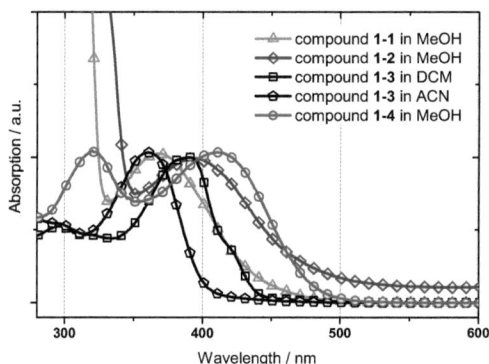

Figure 4.2: UV-visible absorption spectra of the rheniumcarbonyl-complexes 1-1 to 1-4 in different solvents. In more polar solvents such as methanol (MeOH) or acetonitrile (ACN) the lowest lying ^{1}MLCT absorbance is shifted to the red by 30 nm compared to less polar solvents like toluene or dichloromethane (DCM).

strong red shift of the MLCT of about 2770 cm^{-1} as compared to complex 1-1, due to the electron withdrawing effect of the carboxyl groups.

Compound	Solvent	Absorbance	Emission
		[λ_{max}]	[λ_{max}]
(2,2'-bipy.)Re(CO)$_3$Cl (1-1)	Toluene	299, 403	550
(2,2'-bipy.)Re(CO)$_3$Cl (1-1)	Methanol	369	
(4,4'-dicarb.-2,2'-bipy.)Re(CO)$_3$Cl (1-2)	Methanol	394	
(5,5'-bisphen.-2,2'-bipy.)Re(CO)$_3$Cl (1-3)	Dichloromethane	245, 290, 390	442, 689
(5,5'-bisphen.-2,2'-bipy.)Re(CO)$_3$Cl (1-3)	Acetonitrile	360	
([5,5'-bis (LL)-2,2'-bipy.]Re(CO)$_3$Cl (1-4)	Toluene	430	428, 695
([5,5'-bis (LL)-2,2'-bipy.]Re(CO)$_3$Cl (1-4)	Dichloromethane	427	
([5,5'-bis (LL)-2,2'-bipy.]Re(CO)$_3$Cl (1-4)	Methanol	411	

Table 4.1: Summary of photophysical data of rhenium tetracarbonyl diimino complexes 1-1 to 1-4. The abbreviation LL for compound 1-4 is (2,6-bis-octyloxy-4-formyl)phenylethinyl)

The photophysics of the rhenium complex 1-3 is dominated by the presence of the reducing rhenium(I) tricarbonyl chloride donor fragment, which leads to a luminescent lowest-energy triplet excited state of the metal-to-ligand charge transfer type. In some cases, such as with compounds 1-3 and 1-4, the emission quantum yield is very low ($\phi \leq 0.001$). The broad ^{3}MLCT emission of 1-3 in

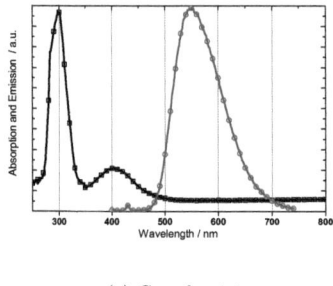
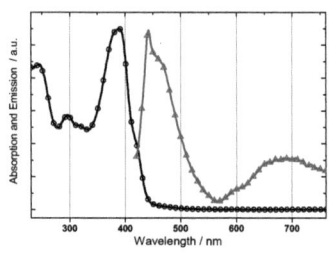

(a) Complex 1-1   (b) Complex 1-3

Figure 4.3: (a) UV-visible absorption (black line with squares) with two maxima at 300 and 400 nm and photoluminescence (red line with triangles) spectra of rhenium carbonyl complex 1-1 in toluene at room temperature and $\lambda_{exc} = 380$ nm. (b) UV-visible absorption (black line with squares) and emission spectra at $\lambda_{exc} = 388$ nm (red line with triangles) of the rhenium carbonyl complex 1-3 in dichloromethane at room temperature. The ^{3}MLCT emission spectra is indicated by a maximum at around 650 nm and additional shoulders at 590 and 700 nm.

dichloromethane solution is also covering a typical wide spectral region including orange and red light exhibiting a maximum at around 650 nm and showing additional shoulders at 590 and 700 nm (Figure 4.3(b)). Interestingly, upon UV-light excitation at the absorption peak maximum of 1-3, the compound also shows an additional structured blue green intraligand emission with maxima at 443 and 465 nm, which is not completely quenched by the lower-lying MLCT states (Figure 4.3(b)).[62] This is not the case for the non-substituted parent compound (2,2'-bipyridyl)Re(CO)$_3$Cl (1-1) (Figure 4.3(a)) and was found to be caused by the additional phenylethinyl groups of the diimine ligand of 1-3.

Comparable dual luminescence behavior has also been reported for similar multichromophore systems investigated by Schanze and coworkers.[64] Additionally compound 1-4, the absorbance and luminescence spectra exhibits similar spectroscopic properties as catalyst 1-3. This is not surprising, considering that 1-3 and 1-4 are structurally related. The IL luminescent transition of 1-4 lies in the range between 400-500 nm in this case, whereas the ^{3}MLCT emission is observed in the region higher than 600 nm.[49]

Chapter 5

Quantum chemical calculations

Computational quantum chemical calculations has developed over the last few decades as a versatile part of science, that generates calculated data which supplements experimental obtained data and helps to determine on the structures, properties and reactions of atoms and molecules. Sometimes quantum chemical calculations allow the determination of properties that are difficult to get experimentally. The quantum mechanical calculations are based primarily on the development of quantum mechanics by the beginning of the 20th century. Nowadays powerful computer programs are used for calculation of electron and charge distributions, molecular geometry in ground and excited states, potential energy surfaces, rate constants for elementary reactions, details of the dynamics of molecular collisions and many other things.[66, 67] In this work quantum mechanical calculations were mainly used for the interpretation of experimental obtained data such as infrared absorption spectra and for the calculation of molecular orbital energy levels.

The method used was similar to the one reported in literature reference nr [62]. The calculations were carried out with Gaussian09.[68] All quantum-chemical calculations were carried out using a density functional theory (DFT) based method with the hybrid B3LYP functional.[69, 70, 71] The 6-31G(d) basis set was used through the calculations,[72, 73, 74] whereas for the complexed rhenium metal the LanL2DZ basis set [75, 76, 77, 78] was applied. The obtained geometries were verified to correspond to a real minimum by establishing an absence of imaginary IR frequencies.

5.1 Infrared absorption spectra

The technique of infrared spectroscopy is a very powerful tool in chemistry and physics. Electromagnetic radiation in the micrometer range, i.e. electromagnetic radiation with wavelengths in the order of 10^{-6} m, has enough energy to excite molecular vibration-rotation transitions.[79] Since molecules have different, characteristic vibration-rotation transitions due to the effective mass of the molecules and force constant of chemical bonds in the molecule, infrared spectroscopy allows for analytical characterization of a molecular substance. However, since the vibrational degree of freedom for non linear molecules scales with $3N$-6, with N being the number of atoms in the molecule, one can imagine that infrared spectra get very complicated for larger molecules making the spectra difficult to interpret. For this reason calculated IR absorption spectra by quantum mechanical DFT calculations are successfully used to correlate characteristic features in the measured spectra to their molecular origin.

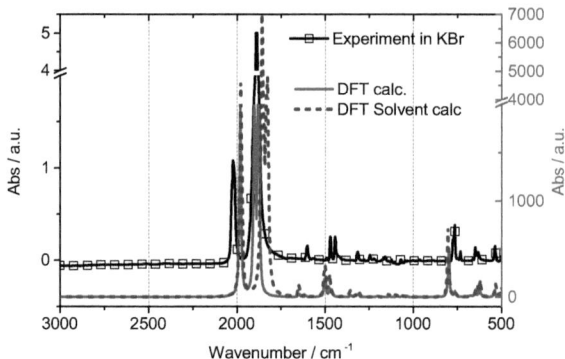

Figure 5.1: IR absorption spectra of Re(2,2'-bipyridyl)(CO)$_3$Cl (1-1). Experimentally measured FTIR difference absorption spectra in KBr (black solid line with squares) and corresponding calculated IR absorption spectra by DFT without (red line) and with (blue dashed line) solvent effects taken into account.

Figure 5.1 shows for example the FTIR difference absorption spectra of compound 1-1 measured in KBr (black solid line with squares) and the corresponding calculated IR absorption spectra by ab initio DFT method (red and blue lines) with the software Gaussian09. It should be noted that Re(2,2'-

CHAPTER 5. QUANTUM CHEMICAL CALCULATIONS

bipyridyl)$(CO)_3$Cl is here used as an example only. The reader will notice that within the thesis many IR-spectra of similar type of materials will be presented and discussed in detail.

The infrared spectrum of complexes of the (LL)Re$(CO)_3$Cl-type are dominated by their characteristic signals connected to the C≡O vibrations positioned at around 1900 cm^{-1} and 2000 cm^{-1} respectively.[46, 62] As the IR absorption bands of the C≡O vibrations are dominant in the spectra and reside in the frequency region where no other absorption bands exist, these bands are useful to probe the electronic exited states of these compounds. Additionally, these band frequencies are greatly influenced by the amount of the electron density on the central rhenium (I) atom, which is considered to be the driving force for catalytic CO_2 reduction.[58] Similar vibrations have been found in for example the related compound (5,5'-bisphenylethynyl-2,2'-bipyridyl)Re$(CO)_3$Cl.[43] The signal at 1650 cm^{-1} is typical for the C=C stretching vibrations in the aromatic ligand system.

5.2. MOLECULAR ORBITAL ENERGY LEVELS

The weak signals around 1500 cm^{-1} are characteristic C-H bending vibrations and those at 880, 800 and 720 cm^{-1} are attributed to C-H rocking vibrations.[80]

As can be seen in Figure 5.1, experimentally observed data and quantum chemical predictions for the IR spectra of the novel compound are in good agreement. For the prediction of infrared spectra it is often necessary to scale the obtained calculated data by a constant factor to achieve good matching of calculated and measured spectra. It was found that calculated ab initio harmonic vibrational frequencies are typically larger than the fundamental vibrational frequencies observed experimentally. Theoretical analysis showed that one main reason of this disagreement is the neglect of anharmonicity effects in the theoretical treatment. Additional complication comes of incomplete incorporation of electron correlation and the fact that used basis sets are finite. Hartree-Fock (HF) theory for example, which is commonly used in DFT based calculations, generally tends to overestimate the calculated vibrational frequencies because of unsuitable dissociation behavior. This limitation can sometimes be overcome by the explicit inclusion of electron correlation. A detailed analysis of this effect with suggestions of different scaling factors can be found in reference [81].

5.2 Molecular orbital energy levels

In this section quantum mechanical calculations were carried out for the determination of molecular orbital frontier energy levels. Figure 5.2 depicts for example the energy levels of the last four occupied and first four unoccupied molecular orbitals (MO) of the compound Re(2,2'-bipyridyl)(CO)$_3$Cl obtained from theoretical calculations at the DFT level.

Similar calculations were carried out for other moleculs of the complexes of the
(LL)Re(CO)$_3$Cl-type. Table 5.1 summarizes the values of quantum mechanical calculations of different rhenium compounds studied in the present work. As can be seen in Table 5.1, quantum mechanical calculations yield a HOMO energy level of about -5.50 eV and a LUMO energy level of about -2.95 eV eV for (5,5'-bisphenylethynyl-2,2'-bipyridyl)Re(CO)$_3$Cl resulting in a band gap of about 2.56 eV. Similar calculations for the simple Re(2,2'-bpy)(CO)$_3$Cl, compare Figure 5.2, gave a lager band gap of 2.89 eV which is in agreement with

CHAPTER 5. QUANTUM CHEMICAL CALCULATIONS

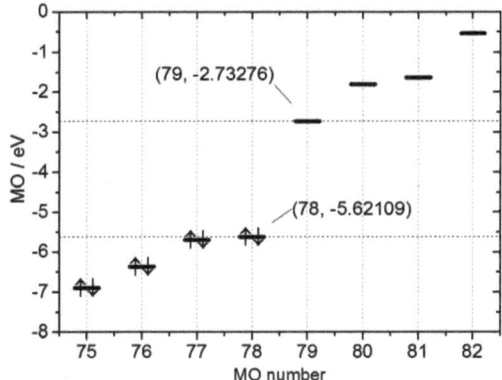

Figure 5.2: Molecular orbital energy levels of Re(2,2'-bipyridyl)(CO)$_3$Cl calculated by DFT for the frontier orbitals including HOMO-LUMO gap.

the UV-Vis absorption measurements depicted in Figure 4.2 in the chapter 4. It should be notice however, that quantum mechanical calculations are based on many assumptions giving only an estimate of the real molecular properties at best.

Compound	HOMO-LUMO [Orbital nr.]	HOMO [eV]	LUMO [eV]	Band Gap [eV]
(2,2'-bipy.)Re(CO)$_3$Cl (1-1)	78 - 79	-5.62	-2.73	2.89
(4,4'-tbut.-2,2'-bipy.)Re(CO)$_3$Cl	110 - 111	-5.45	-2.41	3.04
(4,4'-bisphen.-2,2'-bipy.)Re(CO)$_3$Cl	130 - 131	-5.62	-2.97	2.66
(5,5'-bisphen.-2,2'-bipy.)Re(CO)$_3$Cl	130 - 131	-5.51	-2.95	2.56

Table 5.1: Summary of quantum mechanical calculations of different rhenium compounds studied in the present work.

Many professional programs for quantum mechanical calculations such as Gaussian09 allow impressive visual representation of calculated data. Figure 5.3 shows the visual representation of molecular frontier orbitals of different rhenium compounds as obtained from theoretical calculations at the DFT level. The results obtained from DFT calculations are in particular interesting when compared with optical measurements. The assignment of the lowest-lying excited states of Re(2,2'-bipyridyl)(CO)$_3$Cl and its related compounds in chapter 4 as intraligand (IL) and metal-to-ligand charge transfer (MLCT) type is

5.2. MOLECULAR ORBITAL ENERGY LEVELS

in agreement with the results obtained from the DFT calculations and with previous studies of similar compounds reported in literature.[62] In the visual representation in Figure 5.3 it can be clearly seen, that the calculated highest occupied molecular orbital (HOMO) of the rhenium complex carries a significant metal d-orbital contribution from the $Re(CO)_3Cl$ fragment, which determines the charge transfer character of the lowest excited states in this materials.

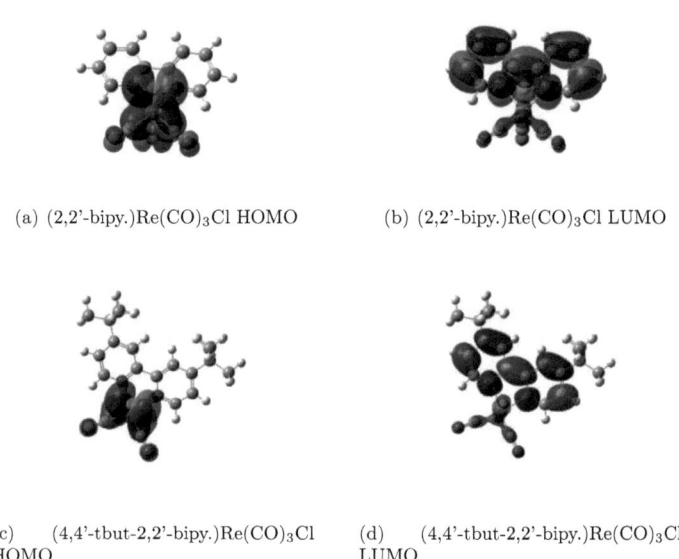

(a) (2,2'-bipy.)$Re(CO)_3Cl$ HOMO

(b) (2,2'-bipy.)$Re(CO)_3Cl$ LUMO

(c) (4,4'-tbut-2,2'-bipy.)$Re(CO)_3Cl$ HOMO

(d) (4,4'-tbut-2,2'-bipy.)$Re(CO)_3Cl$ LUMO

CHAPTER 5. QUANTUM CHEMICAL CALCULATIONS

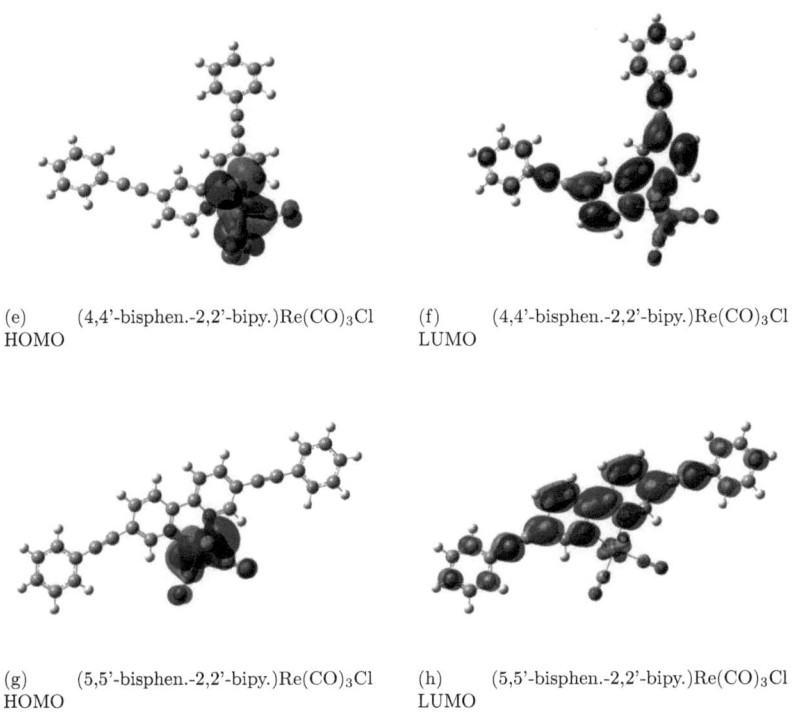

(e) (4,4'-bisphen.-2,2'-bipy.)Re(CO)$_3$Cl
HOMO

(f) (4,4'-bisphen.-2,2'-bipy.)Re(CO)$_3$Cl
LUMO

(g) (5,5'-bisphen.-2,2'-bipy.)Re(CO)$_3$Cl
HOMO

(h) (5,5'-bisphen.-2,2'-bipy.)Re(CO)$_3$Cl
LUMO

Figure 5.3: Visual representation of molecular frontier orbitals of different rhenium compounds studied in the present work.

5.2. MOLECULAR ORBITAL ENERGY LEVELS

Chapter 6

Homogeneous electro catalysis

At present most of the best studied catalysts are metal complexes with bipyridine ligands, where the catalyst center consists of transition metals based on rhenium (Re), rhodium (Rh) or ruthenium (Ru). Despite their high current efficiencies and high selectivity, problems in the field of artificial solar fuel production by these catalysts are manifold. Although these molecular catalyst compounds can be used to stabilize intermediate steps of the CO_2 reduction process and thus lower the required overpotential, achieving a simultaneous multiple electron and proton transfer as indicated by the described reactions 1.3 to 1.8, is kinetically extremely difficult to realize and over potentials of most reported catalyst systems are still significantly high. In Figure 6.1 a schematic representation of a catalyzed and a non catalyzed reaction mechanism with respect to the energy niveau over the reaction coordinate is depicted.

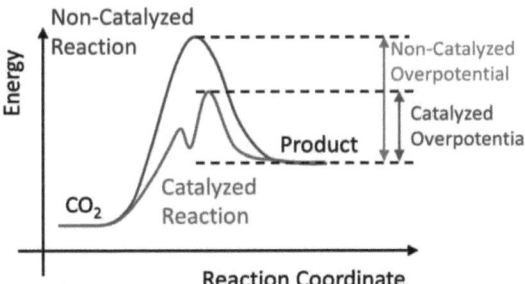

Figure 6.1: Schematic representation of a catalyzed and a non catalyzed reaction mechanism with respect to the energy niveau over the reaction coordinate.

Furthermore, systems based on these catalyst materials, as reported up to now, suffer from low stability and low turnover frequency. In recent yeas Bocarsly et. al. [31] reported on the catalytic reduction of CO_2 to methonol by Pyridinium (the protonated form of Pyridine). Although this system demonstrates only a very low rate of reaction due to a complicated mechanism, which is not jet fully understood, the approach seems to very promising regarding earth abundant catalyst materials and proton rich products. The materials investigated in this chapter dealing with homogeneous electro catalysis are Rhenium diimine complexes with different ligand systems as well as the Pyridinium catalyst. In the following three sections *Rhenium compounds with bipyridine ligands*, *Rhenium compounds with bian ligands* and *Pyridinium as catalyst* the properties of these materials as homogeneous electro catalysts for CO_2 reduction will be investigated.

6.1 Rhenium compounds with 2,2'-bipyridine ligands

Concerning the generation of CO from carbon dioxide, metal complexes with bipyridine ligands are among the most promising candidates for homogeneous catalysis in terms of activities and lifetimes [20, 21]. Up to now, mainly rhenium- and ruthenium-based systems have been reported for their ability to electrochemically or photochemically accelerate the reduction of CO_2 to CO. Carbon monoxide itself can be used as a precursor compound for fuel synthesis processes, where CO and H_2 are mixed as syn-gas to form hydrocarbons such as methane or methanol [25, 82]. Figure 6.2 shows the schematic mechanism for the two electron CO_2 reduction of (2,2'-bipyridyl)Re(CO)$_3$Cl (1-1) and related compounds to CO and CO_3^{2-} as first proposed by Sullivan and Meyer et. al. in 1989.[83]

Scheme 6.3 in this section shows the schematics of four rhenium(I) tricarbonyl chloride complexes with different diimine ligand systems, that is (2,2'-bipy.)Re(CO)$_3$Cl (1-1), (4,4'-dicarboxyl-2,2'-bipy.)Re(CO)$_3$Cl (1-2), (5,5'-bisphen.-2,2'-bipy.)Re(CO)$_3$Cl (1-3) and [5,5'-bis (LL)-2,2'-bipy.]Re(CO)$_3$Cl (1-4) where LL for compound 1-4 is (2,6-bis-octyloxy-4-formyl)phenylethinyl) used for CO_2 reduction.

The (2,2'-bipyridyl) and (4,4'-dicarboxyl-2,2'-bipyridyl) ligand were pur-

CHAPTER 6. HOMOGENEOUS ELECTRO CATALYSIS

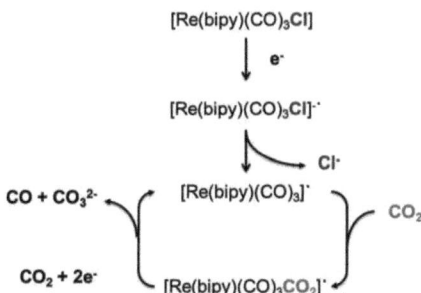

Figure 6.2: Schematic mechanism for the two electron CO_2 reduction of (2,2'-bipyridyl)Re(CO)$_3$Cl (1-1) and related compounds to CO and CO_3^{2-}.

chased from commercial suppliers (Fluka, Aldrich). The multichromophoric diimine ligand (5,5'-bisphen- ylethynyl-2,2'-bipyridyl) was synthesized from 2,2'-bipyridine according to published routes through the formation of 5,5'-dibromo-2,2'-bipyridine and additional Pd-catalyzed Sonogashira coupling of the dibromo-compound with phenylacetylene.[62, 84, 85] [5,5'-bis((2,6-bis-octyloxy-4-formyl)-phenylethinyl)-2,2'-bipyridyl] was obtained via coupling of 5,5'-dibromobipyridine with 1,5-dioctyloxy-4-ethinyl-benzaldehyde. Synthesis of the rhenium complexes was then achieved by further metallation with Re(CO)$_5$Cl in toluene as described in the literature.[54, 86, 87]

Figure 6.4 shows the cyclic voltammograms of 1-1 in nitrogen saturated electrolyte solution on the reductive side (black line with circles) and in a seperate scan the oxidative side (blue line with circles). When the solution was saturated with N_2, compound 1-1 shows a one-electron quasi-reversible reduction wave with its maximum around -1200 mVvs. NHE followed by a one-electron irreversible reduction wave at more negative potential around -1530 mVvs. NHE. This characteristics for rhenium-bipyridine based catalysts are well known and were first reported in this contents by Lehn et al. in 1984.[88] The peak at -1200 mVvs. NHE of compound 1-1 can be attributed to a ligand based reduction and is reversible with its re-oxidation at around -1100 mVvs. NHE. The second reduction wave at -1530 mVvs. NHE can be assigned to a reduction at the metal centre and is not reversible. The oxidative scan (blue line with circles) is somehow interesting since it is usually not reported in most publications where people focus only on the reduction properties of these compounds. On the oxidative side two non reversible oxidation waves are measured. The

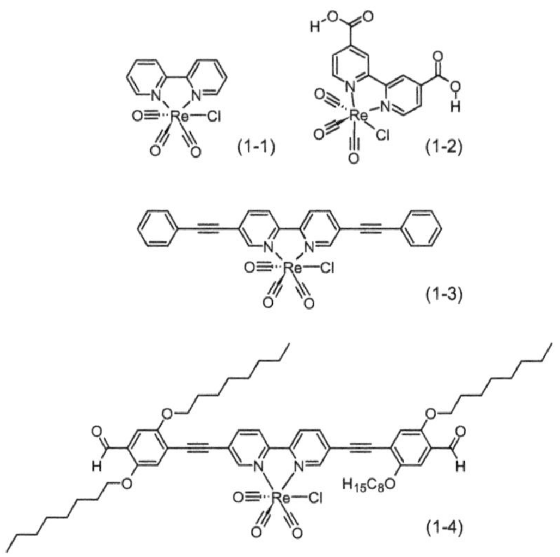

6.1. RHENIUM COMPOUNDS WITH 2,2'-BIPYRIDINE LIGANDS

Figure 6.3: Schematic chemical structures of four different rhenium compounds (2,2'-bipyridyl)Re(CO)$_3$Cl (1-1), (4,4'-dicarboxyl-2,2'-bipyridyl)Re(CO)$_3$Cl (1-2), (5,5'-bisphenylethynyl-2,2'-bipyridyl)Re(CO)$_3$Cl (1-3) and [5,5'-bis ((2,6-bis-octyloxy-4-formyl)phenylethinyl)-2,2'-bipyridyl]Re(CO)$_3$Cl (1-4) for CO_2 reduction.

first with a peak maximum at around 1500 mV vs. NHE and the second with its onset around 1800 mV vs. NHE. The nature of these peaks are speculative and would be a matter for further study. The other compounds presented in this section show a similar behaviour.

Figure 6.5(a) shows a comparison of cyclic voltammograms between compound 1-1 and compound 1-4. These measurements indicate the strong dependence of these compounds on the ligand system. For compound 1-4 the extended conjugated ligand results in a clear shift for the first reduction wave to a more positive potential. As a consequence, also the UV-vis absorption maximum (a MLCT band) is shifted to a longer wavelength as expected and can be seen in Figure 6.5(b). This behavior is important to tune the properties of these metal-organic compounds to improve their capability for electro- and

CHAPTER 6. HOMOGENEOUS ELECTRO CATALYSIS

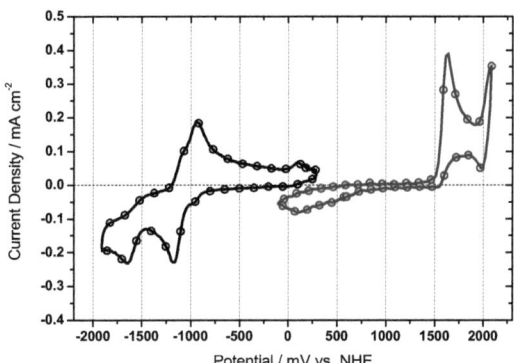

Figure 6.4: Cyclic Voltammograms of 1-1 in nitrogen saturated electrolyte solution on the reductive side (black line with circles) and oxidative side (blue line with circles). Measurements are taken at a scan rate of $100\,\text{mVs}^{-1}$ in acetonitrile with $TBAPF_6$ (0.1 M), Pt working electrode, Pt counter electrode, and a catalyst concentration of 1 mM.

photo- catalytic CO_2 reduction.

Figure 6.6 shows the cyclic voltammograms of 1-1 in nitrogen (black line with squares) and CO_2 (red line with circles) saturated electrolyte solution. When the solution was saturated with N_2, compound 1-1 shows the typical behaviour already described and shown in Figure 6.4. In CO_2-saturated solution (red line with circles), compound 1-1 shows a strong enhancement in current density at the second irreversible reduction wave at about $-1750\,\text{mV}$vs. NHE. This enhancement in current is known to be the catalytic reduction of CO_2 to CO and is assumed to proceed in aprotic solvents according to reaction 1.2. The first quasi-reversible reduction wave at around $-1200\,\text{mV}$vs. NHE does not show a significant increase in current density and is hence not participating in the catalytic reaction directly.

Figure 6.7 shows the same cyclic voltammograms as in Figure 6.6 namely the catalyst 1-1 in nitrogen (black line with squares) and CO_2 (red line with circles) saturated electrolyte solution. Different to the previous measurement however, the working electrode was changed from a platinum to a glassy carbon electrode. Again compound 1-1 shows a strong enhancement in current density at the second irreversible reduction wave at about $-1750\,\text{mV}$vs. NHE indicating

6.1. RHENIUM COMPOUNDS WITH 2,2'-BIPYRIDINE LIGANDS

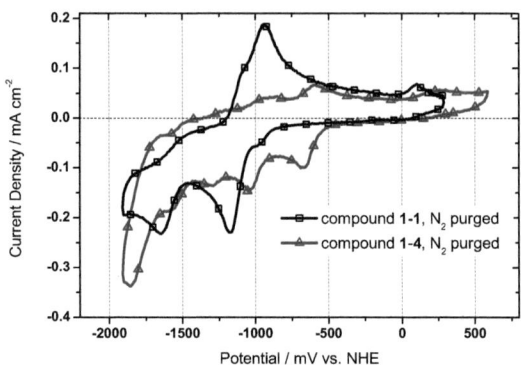

(a) Cyclic voltammograms comparison

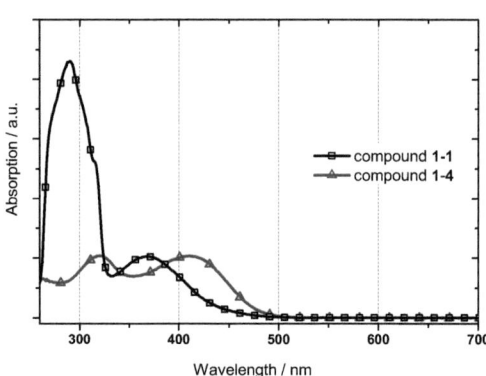

(b) UV-vis absorption spectra comparison

Figure 6.5: (a) Comparison of 1-1 (black line with squares) and 1-4 (blue line with triangles) in cyclic voltammograms and taken at $100\,\mathrm{mVs^{-1}}$ in dimethylformamide with $\mathrm{TBAPF_6}$ (0.1 M), Pt working electrode, Pt counter electrode, and a catalyst concentration of 0.5 mM. (b) Comparison of UV-vis absorption spectra of 1-1 (black line with squares) and 1-4 (blue line with triangles) recorded in methanol.

CHAPTER 6. HOMOGENEOUS ELECTRO CATALYSIS

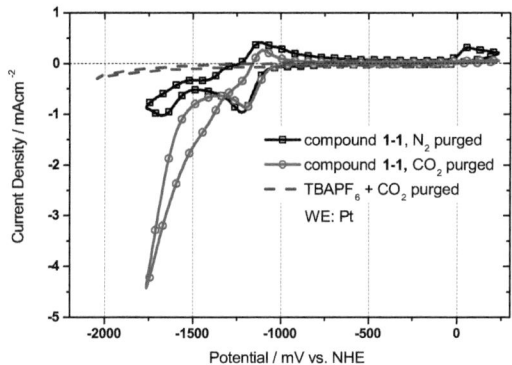

Figure 6.6: Cyclic Voltammograms of 1-1 in nitrogen (black line with squares) and CO_2 (red line with circles) saturated electrolyte solution. Scan with CO_2 saturation shows a large current enhancement due to a catalytic reduction of CO_2 to CO. Measurements are taken at a scan rate of $100\,\mathrm{mVs^{-1}}$ in acetonitrile with $TBAPF_6$ (0.1 M), Pt working electrode, Pt counter electrode, and a catalyst concentration of 1 mM. A scan with no catalyst present under CO_2 (blue dotted line) shows negligible reductive current.

that the catalytic activity of this and similar compounds is not due to the nature of the electrode in use. Additionally this is important due to the fact that CO, which is the main reduction product, is know to lead to a deactivation of the platinum electrode surface by irreversibly binding to and blocking the electrode surface layer.

Although well studied in the past, the catalytic cycle for electrochemical CO_2 reduction of compound 1-1 (and similar compounds presented herein) is still not fully understood. However, it is known that the catalytic mechanism needs an empty coordination site for CO_2 binding which is identified to proceed via the loss of the halide (Cl^-). Some detailed studies on this mechanism have been carried out initially by Sullivan et al.[59] and Lehn et al.[89] in 1985 and 1986, respectively. A more recent study on this was carried out by Johnson et al. in 1996 reviewing several proposed mechanisms.[90]

Figure 6.8 shows the cyclic voltammograms of compound 1-2 measured in N_2 and CO_2 saturated acetonitrile solution, respectively. In comparison to compound 1-1, compound 1-2 does not show a clear and pronounced re-

6.1. RHENIUM COMPOUNDS WITH 2,2'-BIPYRIDINE LIGANDS

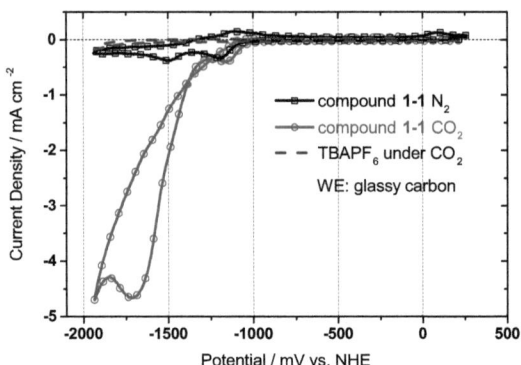

Figure 6.7: Cyclic voltammograms of 1-1 in nitrogen (black line with squares) and CO_2 (red line with circles) saturated electrolyte solution. Scan with CO_2 saturation shows a large current enhancement due to a catalytic reduction of CO_2 to CO. Measurements are taken at a scan rate of $100\,\text{mVs}^{-1}$ in acetonitrile with $TBAPF_6$ (0.1 M), glassy carbon working electrode, Pt counter electrode, and a catalyst concentration of 1 mM. A scan with no catalyst present under CO_2 (blue dotted line) shows negligible reductive current.

versible one-electron quasi-reversible reduction wave around $-1200\,\text{mVvs.NHE}$ although the onset is still observable. A non-reversible wave can be observed at around $-1500\,\text{mVvs.NHE}$ which can be attributed to the additional carboxyl groups on the bipyridyl ligand. In contrast to previously published data,[46] the modified compound 1-2 showed some catalytic behaviour towards CO_2 reduction when the electrolyte solution was saturated with CO_2, as can be seen in Figure 6.8(a) (red line with circles). However, CO_2 potentiostatic bulk electrolysis at $-2100\,\text{mVvs.NHE}$ revealed that the compound seems to be unstable and loses its catalytic activity within several minutes of electrolysis time.

When the acetonitrile solution was saturated with CO_2, also compound 1-3 showed a strong enhancement in the second reduction wave current density. In relative comparison a 6.5-fold increase at the second irreversible reduction wave under CO_2 at $-1750\,\text{mVvs.NHE}$ was observed. Comparing this to compound 1-1, the relative increase in current density is higher for compound 1-3. A detailed study on compound 1-3 has been published.[43]

Figure 6.9(a) shows the cyclic voltammograms of compound 1-3 recorded

CHAPTER 6. HOMOGENEOUS ELECTRO CATALYSIS

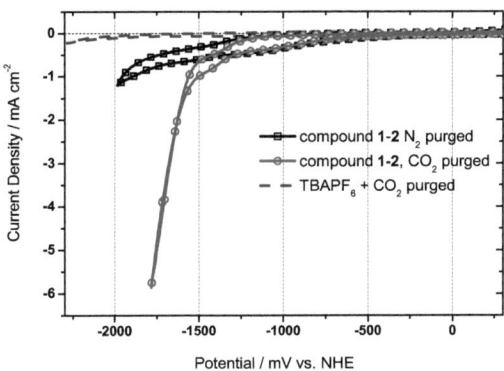

(a) Cyclic Voltammograms of 1-2

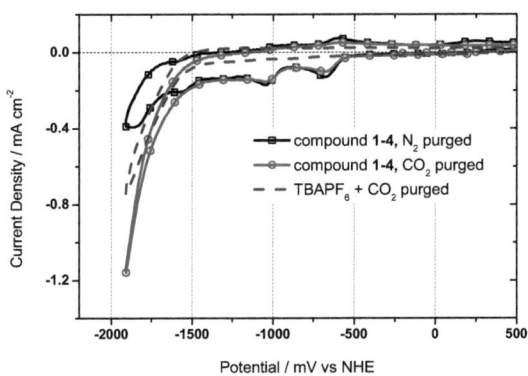

(b) Cyclic Voltammograms of 1-4

Figure 6.8: (a) Cyclic Voltammograms of 1-2 in nitrogen (black line with squares) and CO_2 (red line with circles) saturated electrolyte solution. Scan with CO_2 saturation shows a large current enhancement due to a catalytic reduction of CO_2 to CO. Measurements are taken at a scan rate of $100\,\text{mVs}^{-1}$ in acetonitrile with $TBAPF_6$ (0.1 M), Pt working electrode, Pt counter electrode, and a catalyst concentration of 1 mM. A scan with no catalyst present under CO_2 (blue dotted line) shows negligible reductive current. (b) Cyclic Voltammograms of 1-4 in nitrogen (black line with squares) and CO_2 (red line with circles) saturated electrolyte solution. Scan with CO_2 saturation shows a large current enhancement due to a catalytic reduction of CO_2 to CO. Measurements are taken at the same conditions and concentrations as for compound 1-2.

6.1. RHENIUM COMPOUNDS WITH 2,2'-BIPYRIDINE LIGANDS

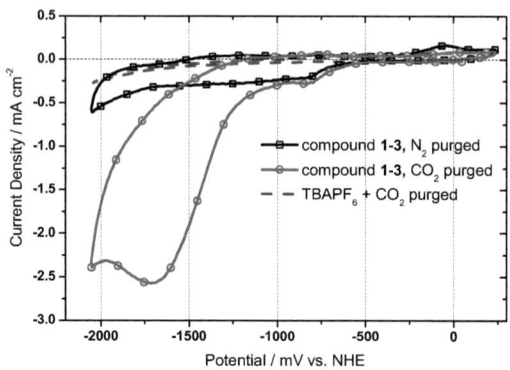

(a) Cyclic Voltammograms of 1-3

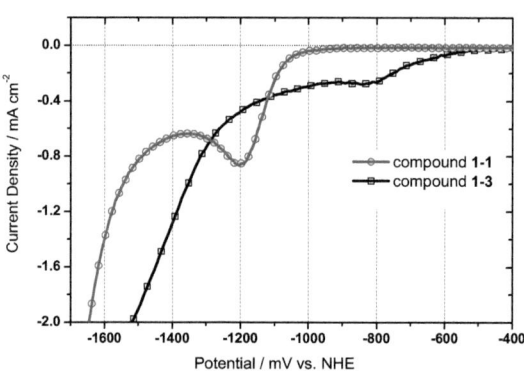

(b) Compound 1-1 compared to compound 1-3

Figure 6.9: (a) Cyclic Voltammograms of 1-3 in nitrogen (black line with squares) and CO_2 (red line with circles) saturated electrolyte solution. Scan with CO_2 saturation shows a large current enhancement due to a catalytic reduction of CO_2 to CO. Measurements are taken at a scan rate of $100\,\text{mVs}^{-1}$ in acetonitrile with TBAPF$_6$ (0.1 M), Pt working electrode, Pt counter electrode, and a catalyst concentration of 1 mM. A scan with no catalyst present under CO_2 (blue dotted line) shows negligible reductive current. (b) Reductive current of compound 1-1 (red line with circles) compared to compound 1-3 (black line with squares) under CO_2 in saturated electrolyte solution. The second reduction wave of compound 1-3 occurs at ca. 300 mV more positive potential compared to compound 1-1.

under N_2- and CO_2-saturated acetonitrile solution. In contrast to compound 1-1, compound 1-3 does not show a clear separation between different reductive waves. Additionally, the onset of the reductive current occurred at a potential around $-750\,mV\,vs.NHE$, which is about 330 mV more positive compared to compound 1-1 (Figure 6.7). This significant differences between compound 1-1 and 1-3 can be attributed to the addition of the bisphenylethynyl groups at the 5,5' position (the phenyl rest is para substituted with respect to the bipyridine ligand). When compound 1-3 was scanned repeatedly under N_2 - saturated conditions to very negative potentials (i.e. $-1600\,mV\,vs.NHE$), a violet film formed on the Pt working electrode. The origin of this film formation and its potential towards CO_2 reduction was further investigated and will be described in great detail in the chapter *Heterogeneous electro catalysis*.

For a detailed comparison of the two compounds 1-1 and 1-3 see Figure 6.9(b). Here only the reductive current of compound 1-1 (red line with circles) compared to compound 1-3 (black line with squares) under CO_2 in saturated electrolyte solution are shown. One can clearly see, that the second reduction wave of compound 1-3 occurs at ca. 300 mV more positive potential compared to compound 1-1.

Different to the compound materials 1-1 to 1-3, compound 1-4 was poorly soluble in acetonitrile. Therefore, cyclic voltammograms were measured in dimethylformamid (DMF). Figure 6.8(b) shows the corresponding cyclic voltammograms of compound 1-4 recorded in N_2- and CO_2-saturated DMF solution. Due to the extended ligand of this compound, the redox characteristics are much more diverse compared to above presented measurements of the more simple compounds 1-1 to 1-3. Within the potential of 0 to $-1800\,mV\,vs.NHE$, 4 distinct reversible reduction peaks can be found with their half-wave potential ($E_{1/2}$) at -635, -1020, -1250 and $-1500\,mV\,vs.NHE$. Since it is generally known for rhenium compounds with bipyridine ligands that a metal-center based reduction shows an irreversible behaviour, these peaks might be attributed to a ligand based reduction. Further studies would be necessary to fully clarify their true nature. Different to the other compounds 1-1 to 1-3 the cyclic voltammogram of compound 1-4 did not reveal catalytic current enhancement within the measured potential range when the electrolyte solution was saturated with CO_2. The recorded current enhancement can be attributed to the additional background current as can be seen from the scan with no catalyst present (blue dotted line).

Figure 6.6 and 6.7 show a crossing of cathodic and anodic currents at around $-1300\,\text{mV}$ vs NHE under CO_2. This trace crossing observed for compound 1-1 indicates the occurrence of a chain process where a species, which is easier reduced than the initial compound, is continuously produced and reduced at the electrode until the trace crossing disappears. This phenomenon has been previously observed and quantitatively interpreted for cyclic voltammetry measurements of electrocatalytic reduction processes.[91, 92, 93]

For the catalyst compound 1-1 one possible explanation for this trace crossing is the formation of a dimer species in lack of sufficient CO_2 present which then can be further reduced to a dimer-anion.[59, 94, 90] These predictions agree with the experimental observations that trace crossing is enhanced with increasing catalyst concentration, compare Figure 6.10(a), and with lower scan rates, compare Figure 6.10(b).

The speed of the CO_2 reduction reaction is an important measure for the capability of a homogeneous catalyst. Comparison between various compounds is generally given by the rate constant k. To define k of the catalyst, usually rotating disk experiments are applied to determine the diffusion coefficient of the complex.[46] For this one can use the Levich equation to calculate the diffusion coefficient according to equation 6.1.[39]

$$i_L = (0.62)nFAD^{2/3}\omega^{1/2}\nu^{-1/6}C \qquad (6.1)$$

Where i_L is the Levich current from the rotating disk experiment, n is the number of electrons, F is the Faraday constant, A is the electrode area, D is the diffusion coefficient, ω is the rotation rate, ν is the kinematic viscosity of the solution and C is the concentration of the analyte in solution.

CHAPTER 6. HOMOGENEOUS ELECTRO CATALYSIS

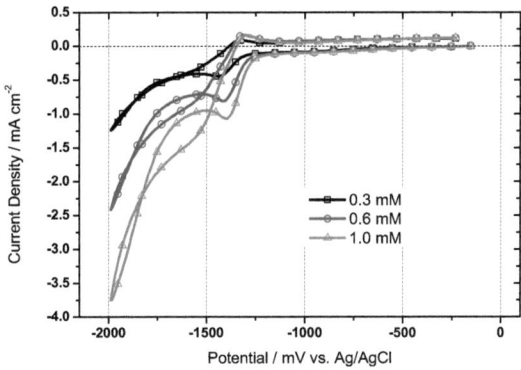

(a) Compound 1-1 at different concentrations

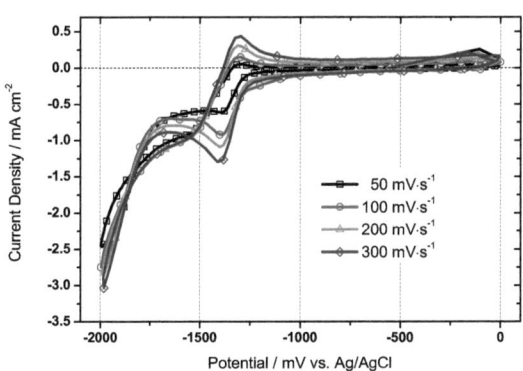

(b) Compound 1-1 at different scan rates

Figure 6.10: (a) Cyclic voltammograms of 1-1 in CO_2 saturated electrolyte solution with different catalyst concentrations. Voltammograms are recorded at $100\,\text{mVs}^{-1}$ using a Pt working electrode, a Pt counter electrode and a Ag/AgCl quasi reference electrode. Trace crossing is enhanced with increasing catalyst concentration. (b) Cyclic voltammograms of 1-1 in CO_2 saturated electrolyte solution at different scan rates. Voltammograms are recorded using a Pt working electrode, a Pt counter electrode and a Ag/AgCl quasi reference electrode. Trace crossing is enhanced with decreasing scan rate.

6.1. RHENIUM COMPOUNDS WITH 2,2'-BIPYRIDINE LIGANDS

The catalytic rate constant (k) can then be extracted from equation 6.2 which holds true for a reversible electron-transfer process followed by a fast catalytic reaction.[95]

$$i_C = nFA[cat]\sqrt{Dk[Q]^y} \qquad (6.2)$$

Where i_C is the catalytic current, $[cat]$ is the catalyst concentration, $[Q]$ is the concentration of the substrate (in this case CO_2), y is the order of the substrate in the reaction in question and the other parameters are the same as in equation 6.1.

However, since rotating disk experiments are usually difficult to obtain, a combination of equation 6.2, for the catalytic current, and equation 6.3, for determining the peak current i_P of a compound with a reversible electron transfer and without any following reaction, can be elegantly used to obtain a good estimate for k simply from cyclic voltammetry measurements.

$$i_P = (0.466)n^{3/2}FA[cat]\sqrt{\frac{DF\nu}{RT}} \qquad (6.3)$$

Where ν is the applied scan rate and the other abbreviations have their described meaning.[39] The ratio of i_C to i_P will then yield equation 6.4, where the diffusion coefficient cancels out and hence k can be estimated knowing i_C and i_P.[96]

$$\frac{i_C}{i_P} = \frac{1}{0.466}\sqrt{\frac{RT}{nF}}\sqrt{\frac{k[Q]^y}{\nu}} \qquad (6.4)$$

The units of k depends on the order of reaction. The homogeneous CO_2 reduction to CO with the compounds 1-1 to 1-4 proceed via a second-order type of reaction resulting in a rate constant k with the units of $M^{-1}s^{-1}$ or L mol^{-1}s^{-1}.[97]

Given the above equation 6.4, detailed analysis of the cyclic voltammogram data as presented in the Figures 6.6 to 6.9(a) allows to estimate the second order rate constant for the CO_2 reduction. Using for example the value of 2.1 for the ratio of i_C to i_P measured for compound 1-1 in equation 6.4 yields in a rate constant of about $60\,M^{-1}s^{-1}$ which is in good agreement to reported literature values.[46] Applying the same method to the other compounds 1-2 and 1-3 gives a rate constant of about $170\,M^{-1}s^{-1}$ for compound 1-2 and a rate constant of about $220\,M^{-1}s^{-1}$ for compound 1-3.

6.2 Rhenium compounds with bis (arylimino) acenaphthene

While for CO_2 reduction up to know mainly polypyridine derivatives have been used as 1,2-diimines, related compounds with chelating imino groups that are not part of a heterocyclic aromatic system were largely neglected. A very attractive example for such a class of ligands are bis(arylimino)acenaphthene derivatives (BIAN-R), which can reversibly store up to four electrons upon reduction and could therefore introduce beneficial effects for accelerating the required multielectron transfer catalysis.[26, 98, 99, 100] In this section the successful application of $Re(BIAN-R)(CO)_3Cl$ complexes as efficient new catalysts for the reduction of carbon dioxide is described.

Scheme 6.11 in this section shows the schematics of three different Re(BIAN-R)$(CO)_3Cl$ compounds (2-1 to 2-3) that have been used as novel materials for the aim of CO_2 reduction. To investigate the redox properties of these compounds 2-1 to 2-3 cyclic voltammograms were recorded using a one compartment cell consisting of a three electrode setup, working, reference and counter electrode. Additionally voltammograms were recorded under CO_2 saturated electrolyte solution to investigate the activities of these compounds towards their capability for CO_2 reduction.

Figure 6.12 shows the cyclic voltammograms of compound 2-3 measured in N_2 saturated acetonitrile solution for 50 and $100\,mVs^{-1}$, respectively. The compound displays four distinct reduction peaks at around -250, -500, -950 and $-1300\,mV$(vs. NHE). The first two are reversible in nature while the last two are partly reversible. The reduction and oxidation of peak 1 and 2 around -250, $-500\,mV$(vs. NHE) displays some characteristic features for a homogeneous,

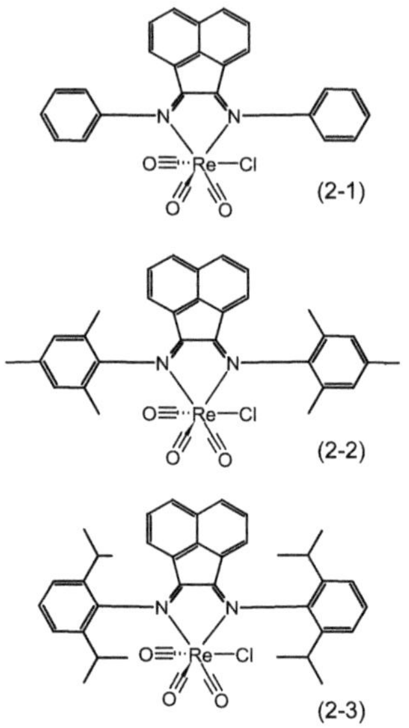

6.2. RHENIUM COMPOUNDS WITH BIS (ARYLIMINO) ACENAPHTHENE

Figure 6.11: Schematic chemical structures of three different rhenium compounds with bis(arylimino)acenaphthene derivatives (BIAN-R) ligands (2-1) (2-2) and (2-3) for CO_2 reduction.

one electron transfer reaction. The peak maxima are separated by approximately 59 mV and the positions of the peak voltage do not change as a function of voltage scan rate. Furthermore the ratio of the peak currents is close to one. Additionally, the peak height scales with a square root dependence on the scan rate suggesting a diffusion controlled process with fast electron transfer as predicted by the Randles–Sevcik equation.[38] The behavior of the last two reduction peaks around -950 and $-1300\,\mathrm{mV}$(vs. NHE) is considerably different.

Figure 6.13(a) shows a comparison of the redox behavior of compound 2-3 between nitrogen and carbon dioxide saturated acetonitrile solution. In carbon dioxide saturated solution (Figure 6.13(a), red curve), compound 2-3 shows a strong enhancement in current density after the 4th irreversible reduction wave at about $-1600\,\mathrm{mV}$(vs.NHE) compared to the situation under N_2 saturation

CHAPTER 6. HOMOGENEOUS ELECTRO CATALYSIS

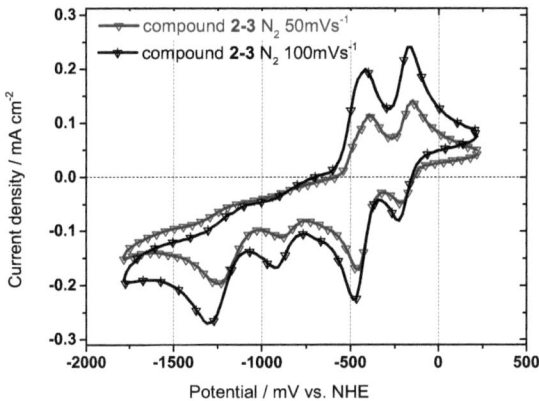

Figure 6.12: Cyclic voltammograms of 2-3 in nitrogen saturated electrolyte solution with two different scan rates of $50\,\text{mVs}^{-1}$ (blue line with triangles) and $100\,\text{mVs}^{-1}$ (black line with slashed triangles) respectively. Measurements are taken in acetonitrile with $TBAPF_6$ (0.1 M), Pt working electrode, Pt counter electrode, and a catalyst concentration of 1 mM.

(Figure 6.13(a), black curve). This enhancement in current is proven to be the catalytic reduction of carbon dioxide to carbon monoxide and is assumed to proceed in aprotic solvents according to reaction 1.2 and in protic solvents according to reaction 1.4. The first four reduction waves at around -250, -500, -950 and $-1300\,\text{mV}$(vs. NHE) do not show a significant increase in current density and are hence not participating in the catalytic reaction directly. The formation of the reduction product CO from CO_2 has been verified after constant potential electrolysis at $-1800\,\text{mV}$(vs. NHE) by GC headspace analysis and additionally with an independent method of gas FTIR absorption measurement as shown in Figure 6.15.

Although well studied in the past for similar Re-based catalysts compounds presented herein, the catalytic cycle for electrochemical carbon dioxide reduction of our new compound 2-3 (and the similar compound 2-2) is still not fully understood and under current investigation. However, it is known that for Re-bipy based catalysts like (2,2'-bipyridyl)$Re(CO)_3Cl$ the catalytic mechanism needs an empty coordination site for carbon dioxide binding, which is identified to proceed via the loss of the halide (Cl^-). For the Re-bipy based system de-

6.2. RHENIUM COMPOUNDS WITH BIS (ARYLIMINO) ACENAPHTHENE

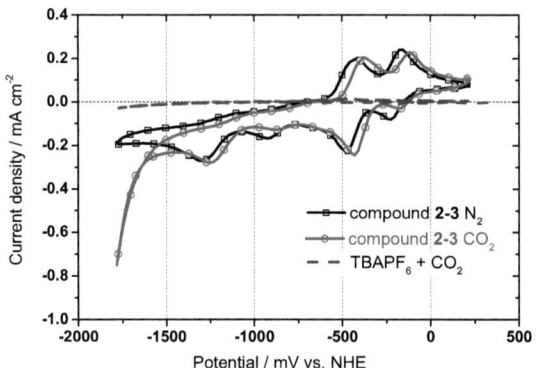

(a) Cyclic Voltammograms of 2-3

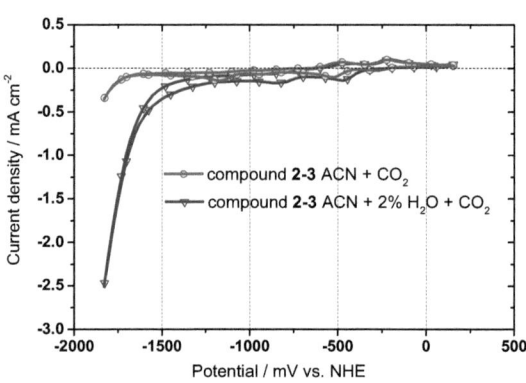

(b) Cyclic Voltammograms of 2-3

Figure 6.13: (a) Cyclic Voltammograms of 2-3 in nitrogen (black line with squares) and CO_2 (red line with circles) saturated electrolyte solution. Scan with CO_2 saturation shows a significant current enhancement due to a catalytic reduction of CO_2 to CO. Measurements are taken at a scan rate of $100\,\mathrm{mVs^{-1}}$ in acetonitrile with $TBAPF_6$ (0.1 M), Pt working electrode, Pt counter electrode, and a catalyst concentration of 1 mM. A scan with no catalyst present under CO_2 (blue dotted line) shows negligible reductive current. (b) Cyclic voltammograms of 2-3 in CO_2 (red line with circles) saturated electrolyte solution and acetonitrile with 2% of H_2O added (blue line with squares). Scan with H_2O added shows a substantial higher reductive current due to an enhanced catalytic reduction of CO_2 to CO and additional H_2O reduction to H_2. Measurements are taken at a scan rate of $50\,\mathrm{mVs^{-1}}$ in acetonitrile with $TBAPF_6$ (0.1 M), glassy carbon working electrode, Pt counter electrode, and a catalyst concentration of 1 mM.

CHAPTER 6. HOMOGENEOUS ELECTRO CATALYSIS

tailed studies on this mechanism have been carried out initially by Sullivan et al.[59] and Lehn et al.[47] in 1985 and 1986, respectively. A more recent study on this was carried out by Johnson et al. in 1996 reviewing several proposed mechanisms.[90] Electrochemical studies and product gas analysis up to know suggest that the catalytic cycle of our new compound 2-3 (and 2-2) proceed in a similar way.

Figure 6.14(a) shows a comparison of the redox behavior of compound 2-2 between nitrogen and carbon dioxide saturated acetonitrile solution. In carbon dioxide saturated solution (6.14(a), red curve), compound 2-2 shows (similar to compound 2-3) a strong enhancement in current density after the 4th irreversible reduction wave at about $-1600\,mV$(vs. NHE) compared to the situation under N_2 saturation (6.14(a), black curve). While otherwise complete similar in its redox behavior compared to compound 2-3 it is interesting to notice that for compound 2-2 the first reduction peak at around $-250\,mV$(vs. NHE) vanishes under CO_2 saturated acetonitrile solution as can be seen in Figure 6.14(a) (red curve). The reason for this is not fully understood up to now.

Figure 6.14(b) shows a comparison of the redox behavior of compound 2-1 between nitrogen and carbon dioxide saturated acetonitrile solution. Different to the other compounds 2-3 and 2-2 presented herein, compound 2-1 does now show any catalytic current enhancement in a carbon dioxide saturated solution (Figure 6.14(b), red curve). Since the compound 2-1 is in structure completely similar to the other compounds 2-3 and 2-2 except its ligand structure on the phenyl rings, the difference in its redox behavior towards its capability of CO_2 reduction can be attributed to this difference on the ligand of the chelating imino groups that are not part of a heterocyclic aromatic system. This would suggest that for any similar type of compound the capability for CO_2 reduction is greatly influenced by its ligand structure.

Figure 6.13(b) shows a comparison of the redox behavior of compound 2-3 when 2 % of H_2O are added to a carbon dioxide saturated acetonitrile solution. In the pure carbon dioxide saturated solution (Figure 6.13(b), red curve), the behavior towards CO_2 reduction is similar to the situation presented in Figure 6.13(a). If water is added to the acetonitrile solution (Figure 6.13(b), blue curve) compound 2-3 shows a substantial further enhancement in current density after the 4th irreversible reduction wave at about $-1600\,mV$(vs. NHE). This enhancement in the reductive current can be attributed to an enhanced

6.2. RHENIUM COMPOUNDS WITH BIS (ARYLIMINO) ACENAPHTHENE

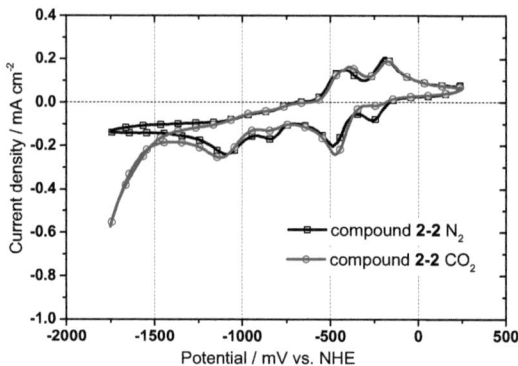

(a) Cyclic Voltammograms of 2-2

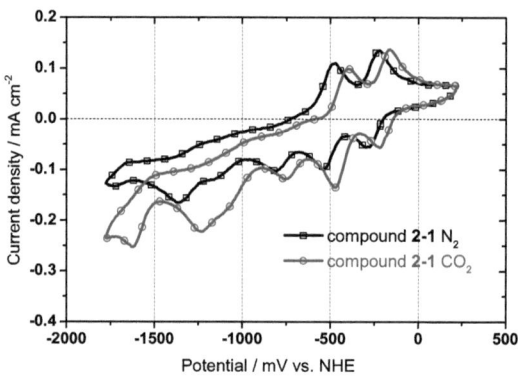

(b) Cyclic Voltammograms of 2-1

Figure 6.14: (a) Cyclic Voltammograms of 2-2 in nitrogen (black line with squares) and CO_2 (red line with circles) saturated electrolyte solution. Scan with CO_2 saturation shows a significant current enhancement due to a catalytic reduction of CO_2 to CO. Measurements are taken at a scan rate of $100\,\text{mVs}^{-1}$ in acetonitrile with $TBAPF_6$ (0.1 M), Pt working electrode, Pt counter electrode, and a catalyst concentration of 1 mM. A scan with no catalyst present under CO_2 (blue dotted line) shows negligible reductive current. (b) Cyclic Voltammograms of 2-1 in nitrogen (black line with squares) and CO_2 (red line with circles) saturated electrolyte solution. Scan with CO_2 saturation shows no significant current enhancement under CO_2 saturation. The measurement conditions and concentrations were otherwise the same as in (a)

CHAPTER 6. HOMOGENEOUS ELECTRO CATALYSIS

catalytic activity towards CO_2 reduction to CO following reaction 1.4 and additionally to the reduction of H_2O to H_2. The formation of both reduction products (CO and H_2) have been verified by GC headspace analysis.

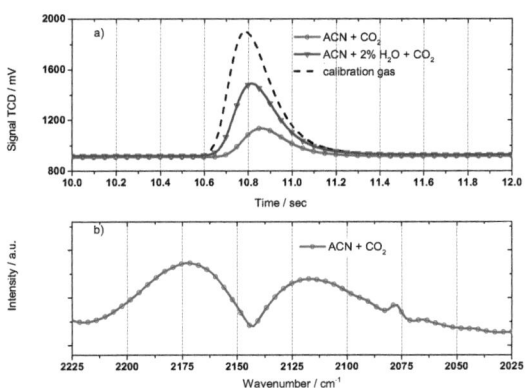

Figure 6.15: Headspace gas analysis after potentiostatic CO_2-electrolysis experiment of 2-3 at constant $-1850\,\mathrm{mV(vs.NHE)}$ (a) GC measurements of the headspace gas for electrolysis experiment performed in acetonitrile solution saturated with CO_2 (red solid line) and acetonitrile solution saturated with CO_2 with 2% H_2O added (blue solid line). For comparison, the calibration gas containing 1 vol% of CO (black dashed line) and (b) FTIR difference absorption spectra in transmission mode of the headspace gas for electrolysis experiment performed in acetonitrile solution saturated with CO_2. (The two peaks centered around $2143\,\mathrm{cm^{-1}}$ correspond to the infrared active vibration of CO)

For a direct proof of the catalytic CO_2 reduction capability of compounds 2-3, headspace gas samples were taken and analyzed regarding the CO concentration by using GC and FTIR as an independent and complementary technique. Figure 6.15 shows the measurements of headspace gas analysis after a potentiostatic CO_2-electrolysis experiment of 1mM 2-3 containing electrolyte solution at constant $-1850\,\mathrm{mV}$(vs. NHE). In Figure 6.15(a) GC measurements of the headspace gas after approximately 10000 seconds electrolysis experiment performed in acetonitrile solution saturated with CO_2 (red solid line) and acetonitrile solution saturated with CO_2 with 2 % H_2O added (blue solid line) are depicted. Additionally, for comparison, a measurement using a standard calibration gas containing 1 vol% of CO (black dashed line) is shown. In Figure 6.15(b) FTIR difference absorption spectra in transmission mode was used to analyze the headspace gas after an electrolysis experiment performed in acetonitrile solution saturated with CO_2. The two peaks centered around $2143\,\mathrm{cm^{-1}}$

correspond to the infrared active rotational-vibrations of the P and R branch of gaseous CO.

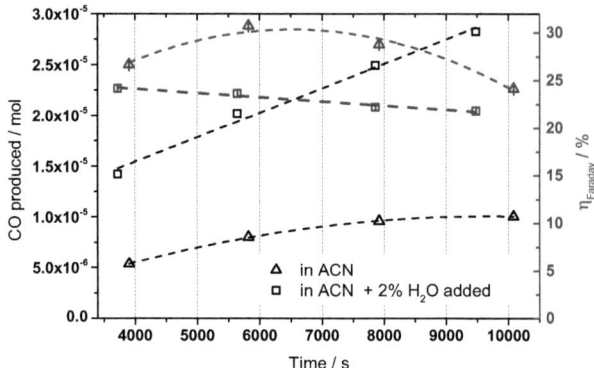

Figure 6.16: Production of CO vs. time plot for potentionstatic CO_2-electrolysis experiment of 2-3 at constant $-1850\,\mathrm{mV(vs.NHE)}$ performed in acetonitrile solution saturated with CO_2 (black triangles) and acetonitrile solution saturated with CO_2 with 2 % H_2O added (black squares) for an electrolysis time of 10000 s. Additionally the corresponding Faradaic efficiencies are shown in blue symbols (triangles for pure ACN and squares for ACN with 2 % H_2O).

For a detailed analysis on the efficiency of the CO_2 reduction process demonstrated by our new compounds 2-3 and 2-2 controlled potential electrolysis were carried out for compound 2-3 and are assumed to be similar for compound 2-2 respectively. Different to the cyclovoltammograms, controlled potential electrolysis experiments were performed using an H-cell setup with separated anode and cathode compartment in order to avoid re-oxidation of the formed products on the counter electrode.

Figure 6.16 shows the production of CO by the catalytic reduction of CO_2 of compound 2-3 over an electrolysis period of approximately 10000 seconds for a pure CO_2 saturated acetonitrile solution (black triangles) and a CO_2 saturated acetonitrile solution containing 2 % of H_2O (black squares). Furthermore, the calculated Faradaic efficiencies for the CO formation are depicted for the water free (blue slashed triangles) and water containing (blue slashed squares) CO_2 electrolysis measurements. In either case the CO production increases over the measurement period of 10000 seconds. However, as already indicated in the cyclovoltammograms of Figure 6.13(b), in the presence of H_2O the formation

CHAPTER 6. HOMOGENEOUS ELECTRO CATALYSIS

of CO is greatly enhanced.

Faradaic efficiencies for the water free system range from around 31 % after 6000 seconds to 24 % after 10000 seconds of electrolysis time. In the water containing system Faradaic efficiencies are noticeably less ranging from approximately 24 % after 4000 seconds electrolysis time to 22 % after 10000 seconds of electrolysis time. This significant difference for the faradaic efficiencies of CO formation can be readily understood by the production of H_2 as competing reaction in the water containing system. A similar behavior is known for the CO_2 reduction to CO of Re-bipy based catalysts like (2,2'-bipyridyl)Re(CO)$_3$Cl. For this type of systems it has been reported that with an addition of 10 % H_2O to the acetonitrile solution a maximum of CO formation can be reached decreasing again with higher amounts of H_2O added.[89]

In summary, the electrocatalytic properties of rhenium(I) tricarbonyl complexes carrying bis(arylimino)acenaphthene (BIAN) ligands have been tested and characterized for the selective two-electron reduction of CO_2 to CO in homogeneous solution. It could also be demonstrated that a variation of the ligand substitution pattern in close proximity to the metal center has a very significant influence on the catalytic performance of these systems. This results have been published in the Journal of *ChemSusChem* compare ref. nr. [101]. Further studies on the suitability of this deeply colored and readily tunable class of compounds[54, 26] as functional components of photocatalytic CO_2-reduction cycles are currently underway.

6.3 Pyridinium and pyridazinium as catalyst

As one more example of homogeneous catalysis the pyridinium-catalyzed reduction of CO_2 to methanol reported in this context first by Bocarsly et.al. [31] is investigated in this section. Bocarsly and his group reported the reduction of carbon dioxide to methanol and formic acid on a platinum electrode in aqueous solutions containing pyridinium ions with faradaic efficiencies up to 20 % according to the mechanism presented in Scheme 6.17.

The study proposed a detailed mechanism of the reduction proceeding through various coordinative interactions between the pyridinium radical and

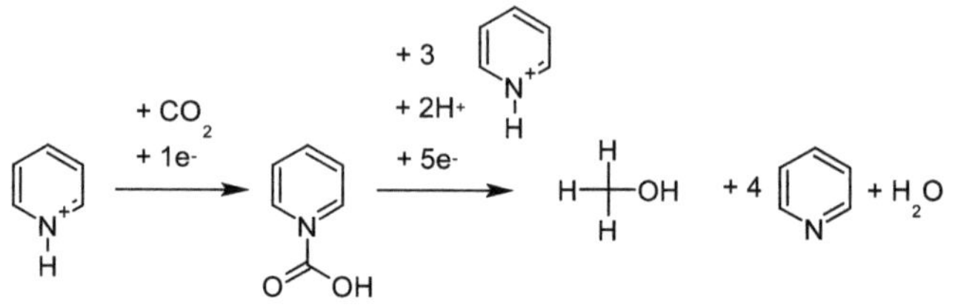

6.3. PYRIDINIUM AND PYRIDAZINIUM AS CATALYST

Figure 6.17: Proposed mechanism for the pyridinium-catalyzed reduction of CO_2 to methanol by Bocarsly et.al. [31]

carbon dioxide, formaldehyde, and additional related species. In the ongoing work several studies have been reported where pyridinium ions were successfully used as catalyst materials towards CO_2 reduction.[102, 103, 104] Additionally a comparative study between pyridine and imidazole was reported which further explored the chemistry of the electrocatalytic reduction of CO_2 using nitrogen containing heteroaromatic systems.[105] Following this reports the proposed mechanism and the role of pyridinium as an active catalyst material lead to an ongoing discussion in chemistry. Especially quantum chemical calculations were carried out to investigate the proposed mechanism and showed that the calculated acidities and redox potentials indicate that pyridinium cations behave differently than previously reported.[106] In a very recent study Saveant et. al. reported that no trace of methanol or formate could be detected upon preparative-scale electrolysis of CO_2 on the same system using pyridinium ions as active catalyst materials.[107]

Following the discussion electrochemical studies using the initially reported system with pyridinium ions as catalyst material in a homogenous aqueous solution and platinum as working electrode material was carried out. The obtained results from cyclic voltammetry studies were compared to a similar heteroaromatic system, pyridazine, containing two nitrogen atoms in the aromatic ring. Although apparently similar in structure, the two systems inherit a strong difference in their pKa value with 5.14 for pyridine and 2.10 for pyridazine.[108]

The different pKa values should result in a significant shift of the acid dissociation equilibrium to regenerate the hydrated protons for the two systems under investigation.

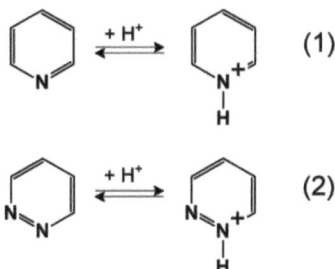

Figure 6.18: Schematic chemical structures of the two different catalyst materials pyridine (1) and pyridazine (2) in pristine and protonated form as pyridinium and pyridazinium

Figure 6.18 shows the schematic chemical structures of the two different catalyst materials pyridine (1) and pyridazine (2) in pristine and protonated form as pyridinium and pyridazinium. Additionally bulk CO_2 electrolysis experiments and product analysis were carried out for both substances. A comparative study for the electrochemical characterization of pyridine and pyridazine and its application towards the reduction of CO_2 to methanol by bulk electrolysis experiments of CO_2 and product analysis by liquid gas and ion chromatography was carried out.

In previous studies mainly NMR spectroscopy with water suppression was used for the detection of methanol as reduction product. Reports showed different results.[31, 107, 33] In this study liquid gas chromatography of the bulk electrolyte solution was used for methanol detection and ion chromatography for the detection of formic acid during bulk electrolysis experiments at constant potential.

Figure 6.19 shows in (a) the cyclic voltammograms of 50 mM pyridine in an aqueous solution of 0.5 M KCl at pH 5.3 and in (b) CVs of 50 mM pyridazine in an aqueous solution of 0.5 M KCl at pH 4.7. recorded under a N_2 atmosphere at a Pt working electrode. The experiments within the potential range between 0 and -850 mV vs. SCE revealed a one quasi-reversible reduction wave centered at -600 mV vs. SCE. The linear dependence of the cathodic and anodic peak current with the square root of the scan rate from 5 to $100\,\mathrm{mVs^{-1}}$

6.3. PYRIDINIUM AND PYRIDAZINIUM AS CATALYST

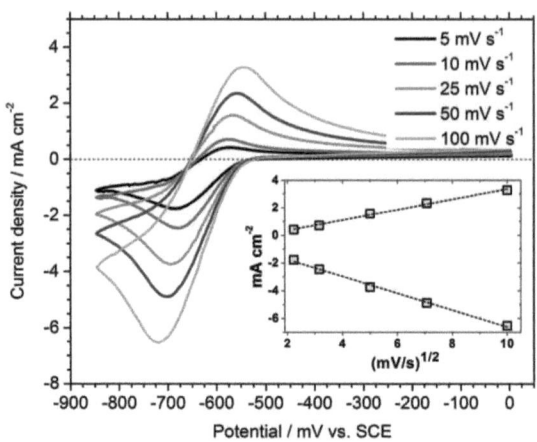

(a) Cyclic voltammograms of 50 mM pyridinium

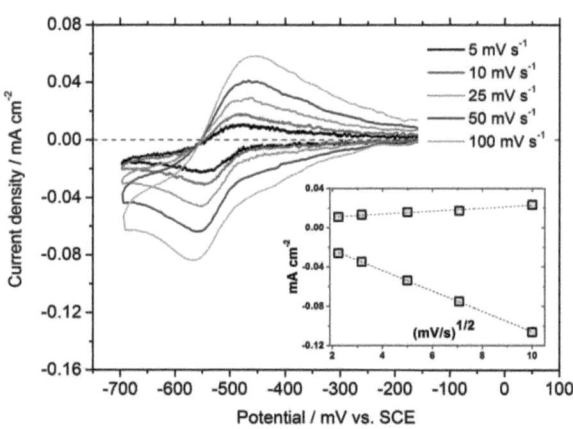

(b) Cyclic voltammograms of 50 mM pyridazinium

Figure 6.19: (a) Cyclic voltammograms of 50 mM pyridine in an aqueous solution of 0.5 M KCl at pH 5.3 and (b) CVs of 50mM pyridazine in an aqueous solution of 0.5 M KCl at pH 4.7. Voltammograms are recorded under a N_2 atmosphere at a Pt working electrode. The scan rates shown are 5, 10, 25, 50, and 100 mVs^{-1}. Inset: Linear dependence of the cathodic and anodic peak current vs. the square root of the scan rate from 5 to 100 mVs^{-1}. [Measurements were done by C. Enengl and S. Enengl]

CHAPTER 6. HOMOGENEOUS ELECTRO CATALYSIS

(R^2 cathodic 0.995, R^2 anodic 0.993), as can be seen in the inset of Figure 6.19(a) and (b), indicates a diffusion limited electrochemical reaction following the Randles-Sevcik equation.[38]

Although both experiments were carried out with identical catalyst concentrations, the electrochemical current response for pyridinium is found to be 75 times higher compared to pyridazinium. This would be in agreement with the mechanism suggested by Saveant et. al. where the reduction of pyridinium is following a reduction of the hydrated protons generated by the rapid dissociation of the pyridinium ions. Since the pKa value of pyridazine is 2.10, the acid dissociation to remain at equilibrium for the regeneration of hydrated protons is less dominant in the cyclic voltammetry measurements, compared to pyridine with a pKa of 5.14.

Figure 6.20 shows the cyclic voltammograms of 50 mM pyridine in an aqueous solution of 0.5 M KCl at pH 5.3 (a) and CVs of 50 mM pyridazine in an aqueous solution of 0.5 M KCl at pH 4.7 (b). The voltammograms are recorded under CO_2 atmosphere for scan rates of 10 and 25 mVs^{-1} respectively. Compared to the situation under N_2 saturation a clear current enhancement in the presents of CO_2 is observed for both systems. For the experiment with pyridinium, the current density increased by a factor of 1.3, from initially -3.8 mA cm^{-2} under N_2 to -5 mA cm^{-2} under CO_2 saturation. For the experiment with pyridazinium, the current density increased by a factor of 5, from initially -0.05 mA cm^{-2} under N_2 to -0.25 mA cm^{-2} under CO_2 saturation. This current enhancement is either attributed to the reduction of CO_2 to methanol by a chain mechanism over several pyridinum radicals, as proposed by Bocarsly et. al.[31] and/or, as Saveant et. al. concluded [107], is simply due to the superposition of the contributions of the two acids present, namely the pyridinium and CO_2 in water.

Similar experiments were carried out with different working electrode materials changing from platinum to glassy carbon, gold and copper, however none of the later showed any noticeable electrochemical response in the applied potential range from 0 to - 800 mV vs. SCE. This characteristic is in agreement with previously reported results showing the important role of platinum in the overall reaction mechanism.[31, 106] In a recent work Musgrave et. al.[109] employed quantum chemical calculations to investigate the role and mechanism of pyridinium based CO_2 reduction. Results indicate a strong binding interaction

6.3. PYRIDINIUM AND PYRIDAZINIUM AS CATALYST

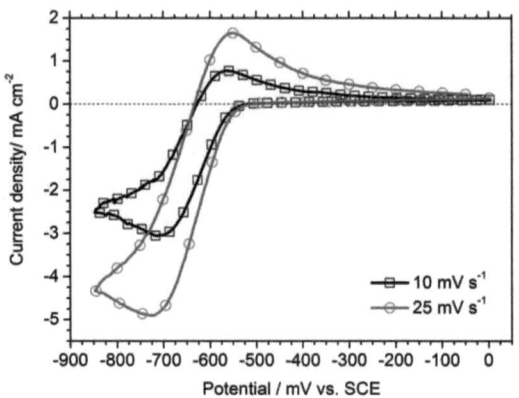

(a) Cyclic voltammograms pyridinium

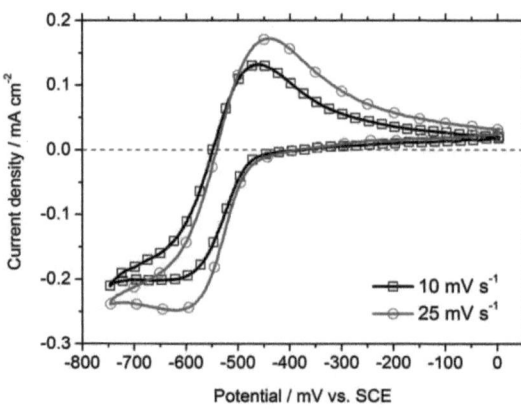

(b) Cyclic voltammograms pyridazinium

Figure 6.20: (a) Cyclic voltammograms of 50 mM pyridinium in an aqueous solution of 0.5 M KCl at pH 5.3 recorded under N_2 (black line with squares) and CO_2 (red line with circles) atmosphere and (b) 50mM pyridazine in an aqueous solution of 0.5 M KCl at pH 4.7. Measurements are recorded at a Pt working electrode and a scan rate of $25\,\text{mVs}^{-1}$. [Measurements were done by C. Enengl and S. Enengl]

CHAPTER 6. HOMOGENEOUS ELECTRO CATALYSIS

of pyridinium with the electrode surface of platinum, resulting in an adsorption energy of 1.0 eV/molecule on Pt(111). It was further concluded that this strong binding interaction of pyridinium with Pt(111) significantly lowers its heterogeneous reduction potential.

Figure 6.21 shows the dependence of the catalytic peak current under CO_2 saturation for different pyridinium and pyridazinium concentrations of 5, 10, 25, 50, 70 and 100 mM respectively. For both catalyst materials the peak current increases with increasing catalyst concentration. For the case of pyridinium as catalyst however, the peak current increases significantly stronger, namely by a factor of 9.2 between a concentration of 5 mM and 100 mM. For the same concentration increase, under otherwise identical conditions, the peak current of the pyridazinium catalyst increases only 1.6 fold, showing that the material is less active towards catalytic CO_2 reduction. When the peak current of pyridazinium under CO_2 saturation is normalized to the peak current under N_2, the peak current ratio decreases with increasing pyridazinium concentration reaching a quasi-constant current ratio of about 3.4 at about 50 mM, compare Figure 6.21(c) in the supplementary information. This is different to the reported behavior of the concentration dependence measured for pyridinium and imidazole, where the current ratio increased, with increasing concentration until the current ratio plateaus. In literature it was concluded that the plateau in current is due to a saturation of active surface-sites.[103, 105] In the case of pyridazinium this characteristic is more indicative of a CO_2 independent increase in base current due to pyridazinium reduction.

A critical experiment to decide on the catalytic reduction of CO_2 is bulk electrolysis followed by product analysis. Several electrolysis experiments for a 50 mM Pyridinum and Pyridazine solution were carried out with initial saturation of the solution by bubbling with CO_2. In both cases controlled potential electrolysis were carried out over an extended period of 30 hours.

Figure 6.22 shows the current-time curve for the constant potential electrolysis experiment of 50 mM pyridine in an aqueous solution of 0.5 M KCl at pH 5.3 (a) and for comparison of 50mM pyridazine in an aqueous solution of 0.5 M KCl at pH 4.7 (b) over an electrolysis time of 16 hours. The data shown in Figure 6.22 correspond to the experimental data shown in the product analysis for methanol depicted in Figure 6.23 in the subsequent information. Current time measurements under CO_2 saturation (red solid line) and N_2 saturation (black

6.3. PYRIDINIUM AND PYRIDAZINIUM AS CATALYST

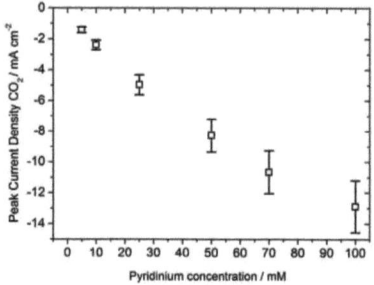

(a) Catalytic peak current under CO_2 for pyridinium

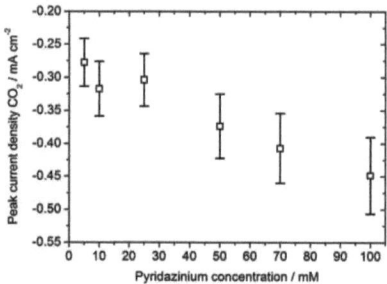

(b) Catalytic peak current under CO_2 for pyridazinium

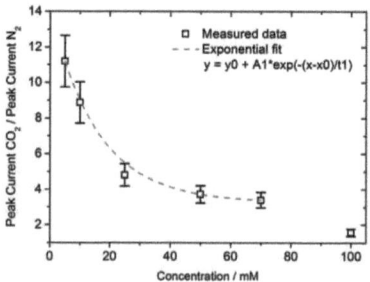

(c) Ratio of current measured under a CO_2 to N_2 for pyridazinium

Figure 6.21: Dependence of the catalytic peak current under CO_2 saturation on different pyridinium and pyridazinium concentrations (5, 10, 25, 50, 70 and 100 mM). All measurements were taken at 100 mV s^{-1} in an aqueous solution of 0.5 M KCl and a pH 5.3 for the measurements with pyridinium (a) and pH 4.7 (b) for the measurements with pyridazinium. (c) Dependence of the catalytic peak current, depicted as the ratio of current measured under a CO_2 atmosphere normalized to the current measured under an N_2 atmosphere in a pH-adjusted solution and increasing pyridazinium concentrations.

CHAPTER 6. HOMOGENEOUS ELECTRO CATALYSIS

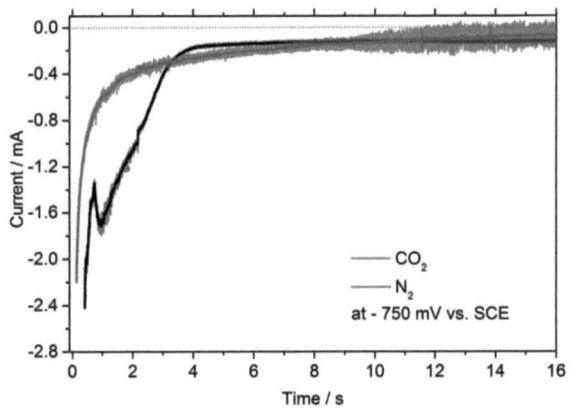

(a) Current-time measurements pyridinium

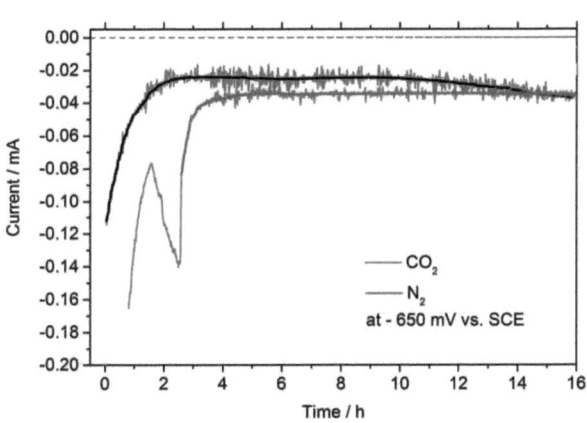

(b) Current-time measurements pyridazinium

Figure 6.22: Current-time measurements for the constant potential electrolysis experiment of 50 mM pyridine at - 750 mV vs. SCE in an aqueous solution of 0.5 M KCl at pH 5.3 (a) and of 50mM pyridazine at - 650 mV vs. SCE in an aqueous solution of 0.5 M KCl at pH 4.7 (b). Current time measurements under CO_2 saturation (red line) and N_2 saturation (black line). [Measurements were done by C. Enengl and S. Enengl]

solid line) are shown for both compounds. The experiment reveals, that there are only minute differences between the current under CO_2 and N_2 saturation for both catalyst materials respectively, which leads to the conclusion that a substantial amount of current is attributed to side reactions despite CO_2 reduction. Such a system is expected to demonstrate only low faradaic efficiencies. This experiment has been repeated for both compounds several times showing similar results. (The initial current increase in the case of pyridinium under N_2 saturation and pyridazinium under CO_2 saturation at the beginning of the measurement is not understood up to know. It is expected to be a characteristic of the experimental setup however, rather than of the catalyst material.) The noise in the current-time characteristics is attributed to the formation of hydrogen bubbles on the working electrode.

Figure 6.23 shows the liquid GC analysis of the electrolyte solution during constant potential electrolysis. The figure on the left side shows measurements of a 50 mM and 10 mM pyridinium concentrated solution in an aqueous 0.5 M KCl at pH 5.3 (a). For both concentrations, samples were taken after 30 hours of electrolysis time at - 800 mV vs. SCE. The measurements show, that for increasing pyridinium concentrations, the production of methanol increases going from 1.7 ppm CH_3OH for the 10 mM concentration to 1.9 ppm for the 50 mM concentration. The corresponding Faraday efficiencies are $9(\pm 1)$ and $14(\pm 1.5)$ % respectively and are hence lower than originally reported Faraday efficiencies for this system of about 22 %.[31] A standard for 12.5 ppm CH_3OH in 0.5 M KCl is also shown in Figure 6.23(a), black solid line.

In comparison on the right side of Figure 6.23, liquid GC analysis of a 50 mM pyridazinium solution in an aqueous 0.5 M KCl at pH 4.7 are shown for 19 and 30 hours of electrolysis time at constant - 650 mV vs. SCE. The corresponding methanol concentrations are 0.2 ppm and 0.3 ppm respectively, corresponding to Faraday efficiencies of about $2(\pm 0.5)$ and $3.6(\pm 0.5)$ %. For comparison a 1 ppm methanol standard in water is also shown (Figure 6.23(b), black line). The retention time for the methanol peak maximum was in all measurements typically at 2.07 min. If one compares this low faradaic efficiencies with the strong current enhancement in the cyclic voltammetry studies between N_2 and CO_2 saturated systems it is noticeable that CO_2 reduction to methanol is only partly responsible for the observed current increase as proposed by Bocarsly et. al.[31]. The additional current increase is expected to come from a

CHAPTER 6. HOMOGENEOUS ELECTRO CATALYSIS

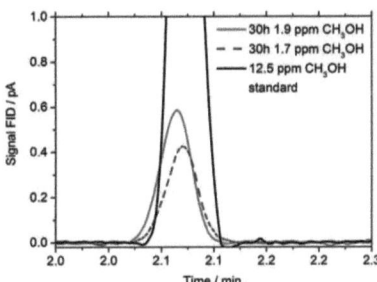

(a) GC analysis for pyridinium electrolysis

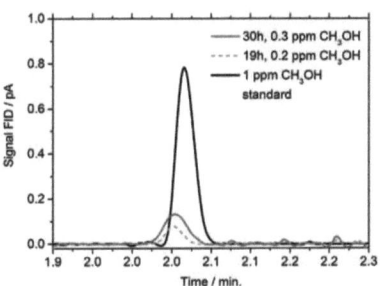

(b) GC analysis for pyridazinium electrolysis

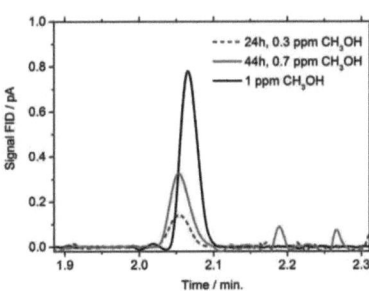

(c) GC analysis for pyridazinium kept at 6°C

Figure 6.23: GC analysis of the electrolyte solution during constant potential electrolysis for pyridinium and pyridazinium. (a) Measurements after 30 hours of CO_2 electrolysis at - 800 mV vs. SCE in an aqueous solution of 0.5 M KCl with 50 (red solid line) and a 10 mM (red dashed line) pyridinium concentration at pH 5.3. (b) GC analysis of 50 mM pyridazine in an aqueous solution of 0.5 M KCl at pH 4.7 for 19 and 30 hours of electrolysis time at constant - 650 mV vs. SCE. (c) Samples from the experiment in (b) taken during electrolysis and kept at 6°C for 24 hours before the measurement

superposition of the contributions of the two acids present, namely the pyridinium and CO_2 in water, or more dominant, the pyridazinium and CO_2 in water, as Saveant et. al. [107] concluded. The difference in the current densities can then be understood by the different pKa values of the two materials with 5.14 for pyridine and 2.10 for pyridazine.

Interestingly it was found that when the samples were stored at 6°C over several hours (e.g. 24 h for the experiment shown in Figure 6.23(c)) and measured again in GC analysis, the methanol concentration increased in all samples significantly and with it, the Faraday efficiency, compare Table 6.1 and Figure 6.23(c) in the presented information. The reason for this increase is not understood by now. Samples that were taken before the electrolysis experiment was started and held under otherwise identical conditions did not yield any methanol signal in the GC analysis. Additionally, the calibration measurement for low methanol concentrations shows excellent linearity from 1 to 50 ppm, compare Figure 2.19(b) in the experimental chapter. A detailed analysis of these measurements and the corresponding calculation of the Faraday efficiencies for methanol formation can be found in Table 6.1. Conclusions

Pyridinium				
Time	Conc. Cat.	Conc. CH_3OH	Coulomb	Faraday eff.
h	mM	ppm	-	%
30	50	1.93	17.3	14.3
30	10	1.79	25.1	9.2
Pyridazinium				
19	50	0.16	3.82	2.4
30	50	0.33	5.29	3.6
Pyridazinium	stored at 6°C			
24	50	0.29	4.57	3.6
28	50	0.33	5.06	3.8
44	50	0.73	6.36	6.6

Table 6.1: Summary of controlled potential electrolysis experiments shown in Figure 6.23 and the corresponding calculated Faraday efficiencies.

In this work the electrocatalytic reduction of CO_2 to methanol and formic acid is explored by the direct comparison of protonated pyridazine and pyridine. Cyclovoltammetric studies for both materials revealed a strong current increase upon CO_2 saturation. The formation of CH_3OH by bulk controlled potential electrolysis experiments could be verified by GC analysis. Faradaic efficiencies

CHAPTER 6. HOMOGENEOUS ELECTRO CATALYSIS

were measured with 14($\pm$1.5) % for the pyrdinium, and 3.6($\pm$0.5) % for the pyridazinium system respectively. The fact that only low faradaic efficiencies were measured although strong current enhancement in cyclic voltammetry studies from N_2 to CO_2 saturated systems were observed lead to the conclusion that CO_2 reduction to methanol is only partly responsible for the observed current increase, as proposed by Bocarsly et. al.[31]. The additional current increase, which is the dominant one, is expected to come from a superposition of the contributions of the two acids present, namely the pyridinium and CO_2 in water, or more dominant, the pyridazinium and CO_2 in water, as Saveant et. al. [107] concluded. This work has been published in the Journal of *ChemElectroChem*, compare ref. nr. [110].

6.3. PYRIDINIUM AND PYRIDAZINIUM AS CATALYST

Chapter 7

Homogeneous photo catalysis

7.1 Results on rhenium-(I) bipyridine complexes

The fac-(2,2'-bipyridyl)Re(CO)$_3$Cl (1-1) is known to be an efficient photo catalyst for the reduction of CO_2 producing mainly CO with the aid of a sacrificial electron donor as for example triethanolamine (TEOA). Figure 7.1 depicts a schematic representation of such a photocatalytic CO_2 reduction system with the catalyst material indecated as (C) and the donor material (D).

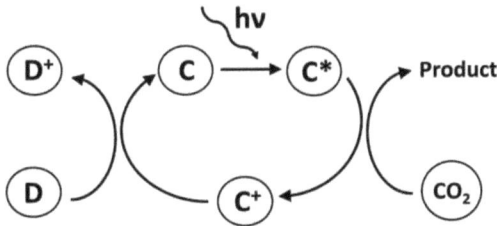

Figure 7.1: Schematic representation of photocatalytic CO_2 reduction with the catalyst material (C) and the donor (D).

The quantum yield (Φ_{CO}) for CO_2 reduction of this compound 1-1 has been reported with 0.14 almost 30 years ago when first published in this context by Lehn et. al. in 1984.[47, 88] Subsequent modifications of the bipyridine ligand system of 1-1 lead to the development of superior catalysts with Φ_{CO} up to 0.59, making these type of materials the most efficient CO_2 photo-catalyst among known homogeneous catalyst materials by now.[58, 111]

7.1. RESULTS ON RHENIUM-(I) BIPYRIDINE COMPLEXES

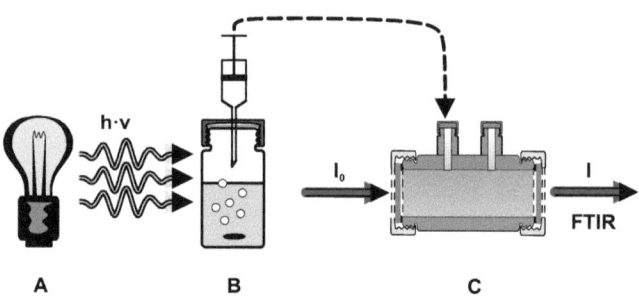

Figure 7.2: Illustration of the photochemical CO_2 reduction experiment with the light source (A), the gas tight reaction cell (B) and the FTIR measurement cell (C).

In this work, the results obtained for the novel compounds (1-3) and (1-5) are compared to our previous findings for (1-1), schematic drawings of all three compounds are depicted in Figure 7.3, showing the modification of additional phenylethynyl substituents at the acceptor ligand 4,4'-position and 5,5'-position. It is expected that this substitution demonstrate an extended conjugation from the bipyridyl acceptor moiety to the central rhenium atom and hence influence the catalyst activity towards electro- and photocatlytic CO_2 reduction.

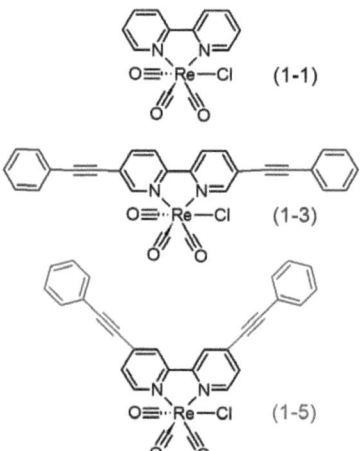

Figure 7.3: Schematic chemical structures of three different rhenium compounds (2,2'-bipyridyl)Re(CO)$_3$Cl (1-1), (5,5'-bisphenylethynyl-2,2'-bipyridyl)Re(CO)$_3$Cl (1-3) and (4,4'-bisphenylethynyl-2,2'-bipyridyl)Re(CO)$_3$Cl (1-5) investigated for photocatalytic CO_2 reduction.

CHAPTER 7. HOMOGENEOUS PHOTO CATALYSIS

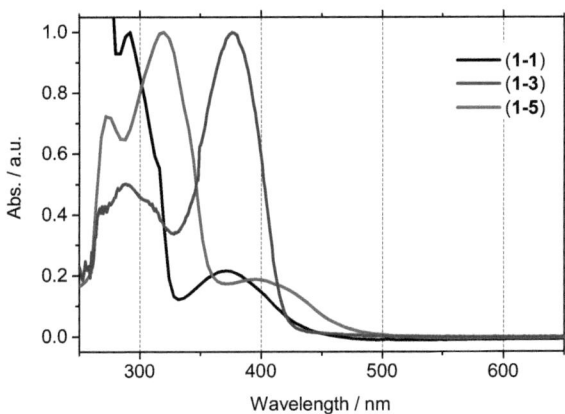

Figure 7.4: Comparison of normalized UV-Visible absorption spectra of three different rheniumcarbonyl-complexes in acetonitrile solution. (2,2'-bipyridyl)Re(CO)$_3$Cl (1-1), (5,5'-bisphenylethynyl-2,2'-bipyridyl)Re(CO)$_3$Cl (1-3) and (4,4'-bisphenylethynyl-2,2'-bipyridyl)Re(CO)$_3$Cl (1-5).

For photocatalysis, it is important to extend the absorption of the catalyst compounds in the visible region. One way to address this is to covalently bind the catalyst to a photosensitizer with high absorption in the visible region. One of the best results concerning quantum yield and turnover numbers was achieved by bridging a Re-based catalyst with a Ru-based photosensitizer reported by Ishitani and coworkers.[60, 112, 113]

A important quantity for photocatalytic systems is the turn over number (TON). It is a dimensionless number and defined as the total number of turnovers the catalyst can achieve until its total decay, independent of the time involved.[114] As such it is an important measure to evaluate the catalyst lifetime and robustness. Experimentally it can be found by the ratio between the amounts of products formed, if there is no limit in substrate (CO$_2$), divided by the amount of catalyst material in the system (mol reduced product of CO$_2$/mol catalyst). Typical TON for Re(bipy)(CO)$_3$X (X=Cl, Br) compounds are in the order of 300.[46, 47, 115]

Scheme 7.2 depicts an illustration of the photochemical CO$_2$ reduction experiment with the light source (A), the gas tight reaction cell (B) and the FTIR measurement cell (C). Gas samples were taken with a gas tight syringe

7.1. RESULTS ON RHENIUM-(I) BIPYRIDINE COMPLEXES

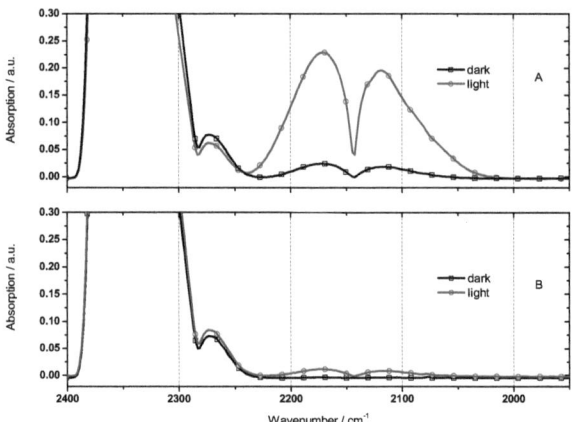

Figure 7.5: IR difference absorption spectra (transmission mode) of headspace samples after illumination (red line with circles) of a DMF:TEOA (5:1/v:v) solution with 2.6 mM catalyst concentration. Control dark experiments are also shown (black line with squares). A, compound 1 and B, compound 2. Light intensity 26900 lux, 360 nm low pass cutoff filter, irradiation time 18 hours.

and transferred to the specially designed FTIR gas cell for recording an IR difference absorption spectrum before and after light irradiation in transmission mode.

Figure 7.5 shows FTIR gas analysis of a photocatalytic reduction experiment of a DMF / TEOA (5:1/v:v) solution containing 2.6 mM of compound 1-1 or 1-3, respectively. The results were compared to a control experiment of the same solution kept in the dark. Headspace samples (5 ml) were taken after an irradiation time of 18 hours and analysed by FTIR spectroscopy. The spectra, as can be seen in Figure 7.5, correspond to a CO gas concentration of A 22.38 vol% and B 0.98 vol% in 5 ml sample gas. The irradiation time was 18 h under 26900 lux with an Osram 400 W Xenophot xenon lamp.

Comparing now photocatalytic CO_2 reduction experiments applying compound 1-1 and 1-3 it was found that the reduction in DMF/TEOA (5:1/v:v), similar to previous works of Koike and coworkers [58], shows big differences in reaction efficiency. As has been reported by other groups before, compound 1-1 showed a good photocatalytic activity with a quantum yield in the range of

$\varphi = 0.167$. In later experiments with the new catalyst 1-3, a significantly lower performance in the generation of CO was reported under similar conditions. In our experiment the new catalyst 1-3 showed a 22.8 fold lesser generation of CO under the same conditions as 1-1. Taking this into account, our best estimate for the quantum yield of catalyst 1-3 (above 360 nm irradiation) is in the order of 0.4 %.[43] These findings can be interpreted by analyzing the above described spectrophotometric measurements, considering the assumption that a long lived ^{3}MLCT* state is a necessary requirement for the successful application of the catalyst in a photochemical system for the reduction on CO_2 to CO.

Although the new catalysts 1-3 and 1-5 show significantly higher absorption in the visible range (compare Figure 7.4), which was assumed to be a clear benefit for photocatalytic application, however, it was shown that the experiments over several hours of irradiation yielded only very low CO formation and under identical conditions, the unmodified compound 1-1 still performs better for photocatalytic CO_2 reduction. To the best of our understanding, following quantum mechanical DFT calculations, this different behaviour might be attributed to an inversion of the lowest-lying excited state properties of compound 1-3 and 1-5 compared to the situation in compound 1-1, which is crucial for the photochemical reactivity of such systems. Such an inversion from the typical metal-to-ligand charge transfer (MLCT) character present in 1-1 to an intraligand (IL) situation was already indicated in a detailed photophysical study of the excited state deactivation pathways of complex 1-3.[62] Further efforts should therefore focus on a systematic tuning of the excited state manifold of compounds such as 1-3 in order to better control the photocatalytic performance while at the same time improving the long-wavelength sensitization of the CO_2 reduction process.[26]

The results on the electrocatalytic performance of the catalysts 1-3 and 1-5 have been published in the Journal of *Electrocatalysis*, compare ref. nr. [116].

7.2 Comparing photo- and electrochemistry

According to the Franck-Condon-Principle, in absorption spectra one always sees the vertical transitions (1) and not the transitions to the relaxation equilibrium state (2), compare Figure 7.6. Therefore what is observed in UV-Vis absorption measurements and what is the actual redox potential is different. This fact also accounts for some of the mismatch between electro- and photochemical measurements.[97, p. 558-565]

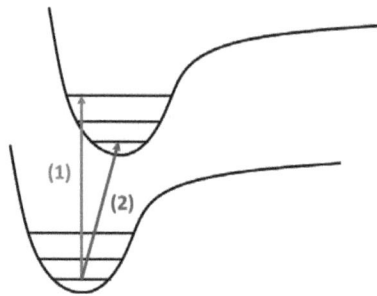

Figure 7.6: Schematic representation of electronic transitions between the electronic ground state and the electronic excited state with vertical transitions (1) and not occuring transitions to the relaxation equilibrium state (2).

The electrochemical one electron reduction of the rhenium-(I) bipyridine (bpy) complex leads to the formation of a $Re^{+1}d^6bipy^{\bullet-}$ radical anion while upon photo excitation one forms a $Re^{+2}d^5bipy^{\bullet-}$ radical anion. Therefore, in photochemistry a sacrificial electron donor (typically TEOA) is used, otherwise the electron would relax to the $Re^{+2}d^5$ state and one would expect the back electron transfer. Since the excited state is relatively long lived, typically several hundred nanoseconds,[53] the $Re^{+2}d^5$ state is quenched by the electron donor and the highest occupied molecular orbital d^5 changes to d^6 and the metal oxidation state from Re^{+2} back to Re^{+1}.

The reason then why the electron transfer is going to the bipyridine-ligand and not to the rhenium metal center can be understood by the use of Marcus theory.[117] According to the Markus theory, the energy of activation $\Delta G^{\ddagger}$ is related to the free energy change of the reaction ΔG^0 and the required reorganization energy λ according to equation 7.1.

$$\Delta G^{\ddagger} = \frac{\lambda}{4}\left(1 + \frac{\Delta G^0}{\lambda}\right)^2 \qquad (7.1)$$

Where the kinetic of the reaction depends on the reorganization energy λ.

CHAPTER 7. HOMOGENEOUS PHOTO CATALYSIS

This becomes more obvious if equation 7.1 is simplified for the case when ΔG^0 is 0 to equation 7.2.

$$\Delta G^{\ddagger} = \frac{\lambda}{4} \qquad (7.2)$$

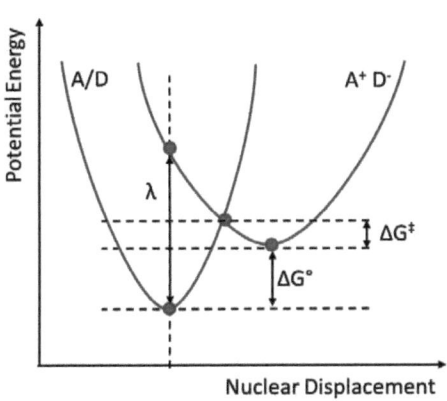

Figure 7.7: Schematic representation of electron transfer from A D to A^+ D^- in a parabola potential energy distribution according to the Marcus theory.

Figure 7.7 shows a schematic visual representation of an electron transfer from species A to D to form A^+ D^- in a parabola potential energy distribution according to Marcus Theory, which is based on quadratic relationships found in parabolas. Additionally, one has to account, that the potential energy for vibrational states of molecules are represented by simple parabolas or harmonic oscillators. In this theory, an electron is transferred from one species to another without change in energy and when an electron is transferred the nuclei can be thought as stationary as the electron transfer occurs in a much faster time scale than nuclear motion, which is also known as the Franck-Condon principle.[97, p. 558-565]

In this context, the rate of electron transfer (k) is related to the activation $\Delta G^{\ddagger}$ by the Arrhenius equation 7.3 .[118]

$$k = A \cdot e^{\frac{-\Delta G^{\ddagger}}{RT}} \qquad (7.3)$$

In the light of Marcus theory, one can now understand, that if the bipyridine ligand is the acceptor of the electron, the negative charge is delocalized over essentially the whole ligand and therefore little reorganization energy λ is necessary. This results in a fast kinetic of the reaction. When the electron is transferred to the rhenium metal center, the electron is much localized and hence the reorganization energy is high. The transfer kinetic becomes slow. The Marcus theory therefore gives the correlation between the rate of the electron transfer and the structural organization of the complex.

Chapter 8

Heterogeneous electro catalysis

While homogeneous catalysis is easier to characterize and mechanistically better understood than heterogeneous catalysis, it also has several disadvantages. High amounts of expensive catalyst material are necessary for efficient CO_2 reduction and the system is limited to the solubility properties of the active species, which often allows only a small variety of different solvents that do not necessarily match desired high CO_2 solubility properties. Furthermore homogeneous catalysts may sometimes face solution deactivation pathways, as for example a dimer formation of Re-compounds with bipyridine ligands in non-aqueous solution systems.[59] One way to overcome these problems is to immobilize the catalyst on the electrode and thereby change from homogeneous to heterogeneous catalysis. In the past, the most frequently reported ways to immobilize Re type catalysts onto a solid electrode were either the insertion of the molecule in a polymer matrix [89, 119, 120] or the chemical modification of the ligand-molecule with a functional group allowing polymerization.

8.1 (2,2'-bipy.)Re(CO)$_3$Cl incorporation into a polymer matrix

A versatile technique for the immobilization of an active catalyst species is the incorporation of the catalyst material in a host polymer matrix. Yoshida et. al. successfully incorporated the (2,2'-bipyridyl)Re(CO)$_3$Br and (terpyridine)-Re(CO)$_3$Br into a coated Nafion membrane for the reduction of CO_2 to formic

8.1. (2,2'-BIPY.)RE(CO)₃CL INCORPORATION INTO A POLYMER MATRIX

acid and CO in a water based environment.[121]

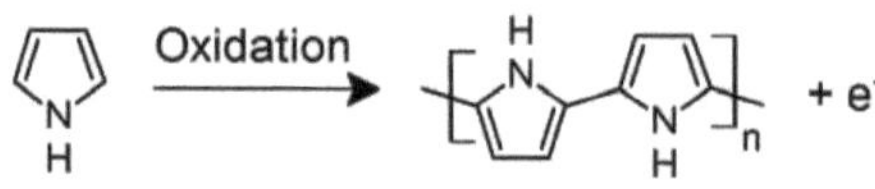

Figure 8.1: Schematic representation of the electrochemical polymerization of pyrrole to polypyrrole uppon oxidation.

In our approach the aim was to immobilize the catalyst material (2,2'-bipyridyl) $Re(CO)_3Cl$ (1-1) into a polypyrrole matrix by electrochemical polymerization of pyrrole in a homogeneous mixture of pyrrole and the catalyst material. Figure 8.1 shows a schematic representation of the electrochemical polymerization of pyrrole to polypyrrole by oxidation. Pyrrole was electropolymerized on a Pt foil serving as supporting working electrode for the polypyrrole film. Pyrrole was used as received. 625 µl of pyrrole were added to 18 ml of acetonitrile with $TBAPF_6$ (0.1 M) as supporting electrolyte to receive a monomer concentration of 0.5 M. The electropolymerisation was performed by sweeping the potential between 1053 mV and -647 mV vs. NHE over 70 cycles with a scan rate of $100\,\text{mVs}^{-1}$.

Figure 8.2(a) shows the potentiodynamic film formation of a pure polypyrrole film without compound 1-1 present. The successive film formation is indicated by an increase in the current density of each cycle. The film formation ceased after approximately 70 cycles when no additional current increase could be observed. After electropolymerization the Pt working electrode was fully covered with a continuous dark black film of polypyrrole, compare Figure 8.2(c). In Figure 8.2(b) the pure polypyrrole electrode on Pt was tested upon CO_2 reduction. Cyclic Voltammograms of a polypyrrole covered Pt electrode in N_2 (black line with squares) and CO_2 (red line with circles) saturated electrolyte solution are shown. The scan with CO_2 saturation shows no significant reductive current enhancement.

In a subsequent similar experiment compound 1-1 was dissolved in the pyrrole monomer electrolyte solution with a concentration of 2 mM and used for electro polymerization. After electropoylmerization at constant current of 0.1 mA the formed film was removed from the system and washed with pure acetonitrile solution to remove any initial monomer and catalyst material not

CHAPTER 8. HETEROGENEOUS ELECTRO CATALYSIS

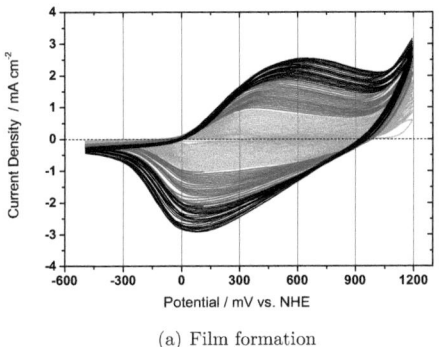

(a) Film formation

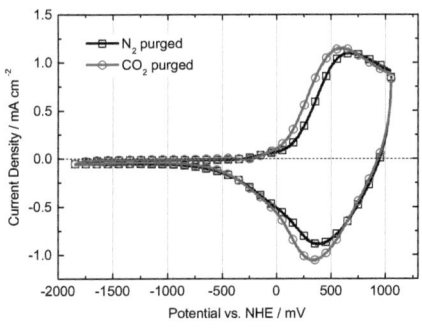

(b) Film characterization

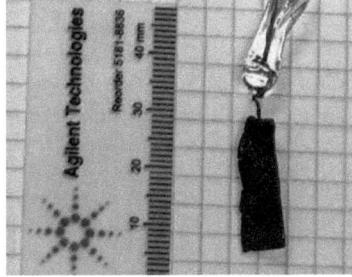

(c) Polypyrrole film on Pt

Figure 8.2: (a) Potentiodynamic electropolymerisation of pure Pyrrol over 70 cycles on a Pt foil by sweeping the potential between 1053 mV and -647 mVvs. NHE over 70 cycles with a scan rate of $100\,\mathrm{mVs^{-1}}$. With ongoing film formation the current density increases up to a maximum until the film formation ceased. (b) Film characterization at $50\,\mathrm{mVs^{-1}}$ in N_2 (black line with squares) and CO_2 (red line with circles) saturated electrolyte solution. (c) Picture of the electropolymerized polypyrrole film on a Pt foil supported working electrode. [Measurements were done by S. Schlager]

8.1. (2,2'-BIPY.)RE(CO)$_3$CL INCORPORATION INTO A POLYMER MATRIX

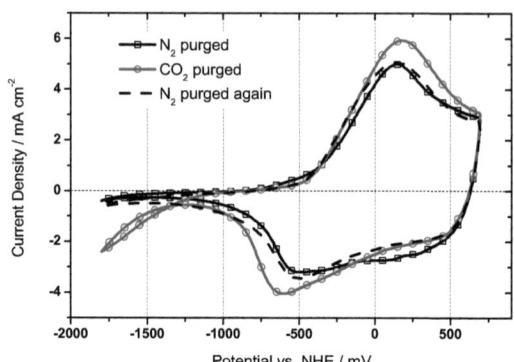

Figure 8.3: Cyclic Voltammograms of a polypyrole covered Pt electrode and incoporated catalyst compound 1-1 in N_2 (black line with squares) and CO_2 (red line with circles) saturated electrolyte solution. Scan with CO_2 saturation shows a significant reductive current enhancement attributed to a catalytic reduction of CO_2 to CO by the catalyst. Measurements are taken at a scan rate of $100\,\mathrm{mVs^{-1}}$ in acetonitrile with TBAPF$_6$ (0.1 M) and a Pt counter electrode. Upon sequential N_2 purging the original cyclic voltammetry curve is restored (black dashed line).

incorporated into the polymer matrix. Then the polymer film was used as working electrode for heterogeneous CO_2 reduction.

Figure 8.3 shows the cyclic voltammograms of the polypyrrole covered Pt electrode and the incorporated catalyst compound 1-1 in N_2 (black line with squares) and CO_2 (red line with circles) saturated electrolyte solution. The scan with CO_2 saturation shows a significant reductive current enhancement which is assumed to be due to a catalytic reduction of CO_2 to CO by the catalyst 1-1. Measurements are taken at a scan rate of $100\,\mathrm{mVs^{-1}}$ in acetonitrile with TBAPF$_6$ (0.1 M) and a Pt counter electrode. Upon sequential N_2 purging the original cyclic voltammetry curve is restored (black dashed line). The reduction peak at about $-600\,\mathrm{mV}$ vs. NHE is not present in the cyclic voltammogram of the pristine film depicted in Figure 8.2(b) before and is hence attributed to the first, bipyridyl based reduction of the catalyst 1-1, compare Figure 6.6 in chapter 6.

Similar results have been reported by Chen et. al. where Cobalt-Phthalocyanine was successfully incorporated into a polypyrrole matrix for CO_2 reduction.[122]

It has to be pointed out, that in this experiments no product gas analysis was performed to independently verify the CO_2 reduction and confirm the expected CO formation. Subsequent experiments failed in reproducing this data and further studies on this effect are under current investigations.

8.2 (5,5'-bisphen.-2,2'-bipy.)Re(CO)$_3$Cl polymerization

This chapter investigates the electropolymerization of compound 1-3 onto the electrode and determines its potential for heterogeneous catalysis towards CO_2 reduction. The film growth on a Pt working electrode was performed by potentiodynamic reductive scanning in nitrogen saturated acetonitrile solution containing 0.1 M TBAPF$_6$ and the catalyst monomer with a concentration of 2 mM. The films were electrochemically characterized using cyclic voltammetry. As done before, the catalytic properties for CO_2 reduction were studied via cyclic voltammetry in carbon dioxide saturated acetonitrile solution containing also 0.1 M TBAPF$_6$.

Figure 8.5 shows the potentiodynamic formation of the rhenium catalyst film by electro-polymerization of the rhenium catalyst monomer 1-3 on a Pt working electrode from a 2 mM monomer solution. Changes in the voltammogram with increasing number of cycles are indicated by the numbers 1 to 5 in Figure 8.5. In the first scan (red line with circles), two distinct reduction waves of the monomer 1-3 are still visible. The peak at (−850 mVvs.NHE) can be attributed to a ligand based reduction, and the reduction wave at (−1300 mVvs.NHE) can be assigned to a reduction at the metal centre. The first reduction wave is partly reversible and the re-oxidation peak appears at (−750 mVvs.NHE) . Both peaks 1 and 4, attributed to the monomer ligand, decrease with increasing number of scans, which suggests a ligand-based polymerisation of the monomer 1-3. The polymerization is assumed to proceed via radical coupling between two electro-generated radical species as was reported previously for similar systems,[47, 123] although alternative routes cannot be excluded due to the presence of metal carbonyl species.[124] Furthermore, the maximum of the first reduction wave 1 shifts towards more negative potentials with increasing number of scans. An oxidative peak 5 at around (−100 mVvs.NHE) initially appears and disappears after continuous

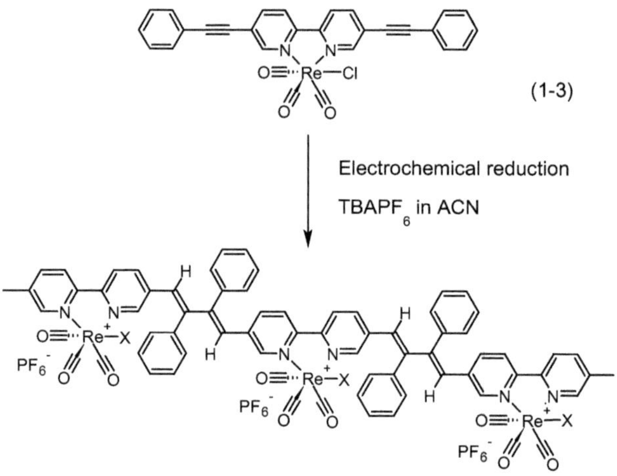

8.2. (5,5'-BISPHEN.-2,2'-BIPY.)RE(CO)$_3$CL POLYMERIZATION

Figure 8.4: Schematic representation of the electrochemical polymerization of the monomer compound (1-3) and the resulting possible chemical substructure of the rhenium sites within the polymer film in which X represents a chloride or a substituted ligand from the reaction medium.

scanning which may be attributed to a temporary dimer formation as it is described for similar systems. After approximately 25 cycles, the voltammogram doesn't change any further and shows a distinct background current below (-700 mVvs.NHE) (blue line with triangles). This background current is present over continuous scans and indicated by 2 and 3. After formation, the polymer film shows an intense violet colour on the part of the electrode that was in contact with the monomer solution (compare Figure 8.9).

Figure 8.6 shows the electroactivity of the rhenium catalyst film on a platinum plate electrode at various scan rates from 200 mVs^{-1} (blue line with triangles) to 10 mVs^{-1} (red line with circles). A plot of peak current vs. scan rate reveals a linear dependence suggesting that the redox process is not any more diffusion controlled as predicted by the Randles-Sevcik equation.[39] This further confirms the formation of an electroactive film immobilized on the Pt electrode surface.[125] Additionally, the maximum reduction peak position is independent of the scan rate within the measured cycling times. This shows

CHAPTER 8. HETEROGENEOUS ELECTRO CATALYSIS

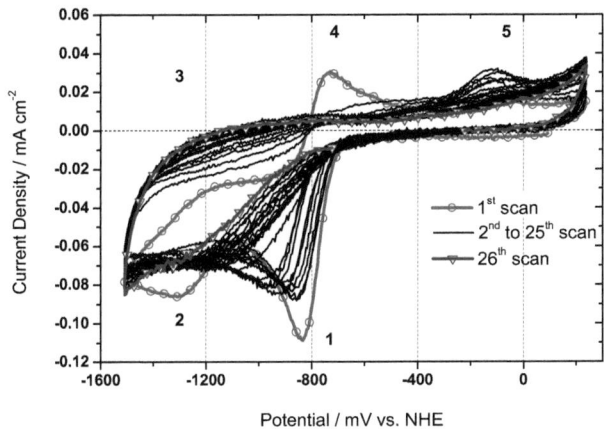

Figure 8.5: Potentiodynamic formation of rhenium catalyst film on Pt from a catalyst monomer solution of 1-3. First scan (red line with circles) and last scan (blue line with triangles). Voltammograms are recorded at $50\,\text{mVs}^{-1}$ in nitrogen-saturated acetonitrile solution containing 0.1 M TBAPF$_6$ and a monomer catalyst concentration of 2 mM.

that the electron transfer kinetics is fast with respect to the cycling time scales suggesting a Nernst like behaviour.[38]

Figure 8.7 shows cyclic voltammetry measurements of the rhenium catalyst film on a Pt plate electrode in N_2- and CO_2-saturated electrolyte solution. The measurement of the potential window with two Pt electrodes as WE and CE, respectively, and an electrolyte solution under N_2 does not show any reductive current in the potential range from 0 mVvs.NHE to -2000 mVvs.NHE. When the solution is purged with CO_2 for 10 min and no catalyst is present, a reductive current starts to flow at a potential lower than about -1700 mVvs.NHE (blue dashed line). When the Pt working electrode is replaced by the Pt electrode with the rhenium catalyst film and measured in the electrolyte solution under N_2 atmosphere, the typical reduction curve as shown in Figure 8.6 is measured again (Figure 8.6, black line with squares). If however the electrolyte solution is at CO_2 saturation, a high non-reversible reductive current enhancement is observed (Figure 8.6, red line with circles). The reductive current begins to increase at about -1150 mVvs.NHE and can be attributed to the reduction of CO_2 to CO.

8.2. (5,5'-BISPHEN.-2,2'-BIPY.)RE(CO)$_3$CL POLYMERIZATION

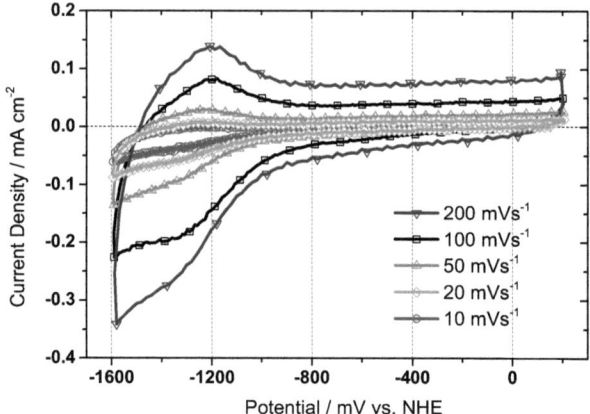

Figure 8.6: Cyclic voltammograms of the rhenium catalyst film on Pt in nitrogen saturated acetonitrile solution containing 0.1 M TBAPF$_6$ at different scan rates from 200 mVs^{-1} (blue line with triangles) 10 mVs^{-1} (red line with circles).

According to previous studies on active electrodes with rhenium based catalysts, the pathway for CO$_2$ reduction is similar to the homogeneous system. The catalytic mechanism proceeds via coordination of a CO$_2$ molecule to a rhenium atom, which allows the reduction of CO$_2$ to CO.[59, 123, 83, 126, 127] As a result, the catalytic current per area depends on the number of active redox sites per surface area.

A comparison between the catalyst monomer 1-3 in solution (green line with triangles) and the polymerized rhenium catalyst film on a Pt plate electrode (red line with circles) in CO$_2$-saturated electrolyte is presented in Figure 8.8. The measurement shows that the electrochemical potential onset for the CO$_2$ reduction with the rhenium catalyst film on the Pt electrode has a similar value as the onset for CO$_2$ reduction using the monomer substance 1-3 in solution, which is at about -1150 mVvs.NHE. The reductive current increases initially more rapidly for the homogeneous monomer system. With increasing negative potential, however, the current density at the catalyst film electrode increases significantly faster, surpassing the reductive current of the 1 mM monomer solution at about -1450 mVvs.NHE.

CHAPTER 8. HETEROGENEOUS ELECTRO CATALYSIS

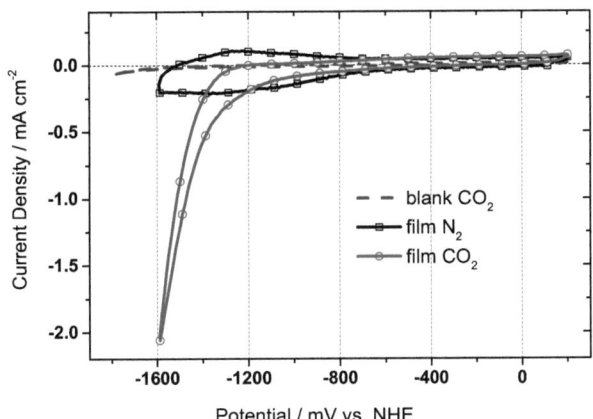

Figure 8.7: Cyclic voltammograms of the rhenium catalyst film on a Pt plate electrode in nitrogen- (black line with squares) and CO_2- saturated electrolyte solution (red line with circles), respectively. The scan in the presence of CO_2 shows a large current enhancement due to the catalytic reduction of CO_2 to CO. A scan with no catalyst film present under CO_2 (blue dashed line) shows little to no reductive current. Voltammograms were recorded at $100\,mVs^{-1}$ in acetonitrile with a Pt counter electrode.

In contrast to the cyclic voltammogram of the rhenium catalyst film on a Pt electrode (red line with circles), the cyclic voltammogram of the catalyst monomer 1-3 in a CO_2- saturated electrolyte solution (green line with triangles) still shows the quasireversible first reduction wave at about $-850\,mV$vs.NHE. This reduction wave is attributed to the ligand of the monomer 1-3. As known from previous experiments, this reductive peak does not show any current enhancement under CO_2- saturation compared to saturation under N_2.[83, 43]

CO_2-electrolysis experiments at constant $-1600\,mV$vs.NHE of a pure platinum plate electrode and of the rhenium catalyst film was performed in acetonitrile solution saturated with CO_2. The experiment was carried out in a sealed cell over 60 minutes of electrolysis time. During this period no film degradation was observed. Current-time plots can be found in the *Experimental techniques* section compare Figure 2.9.

As a direct proof of the catalytic CO_2-reduction capability of the rhenium

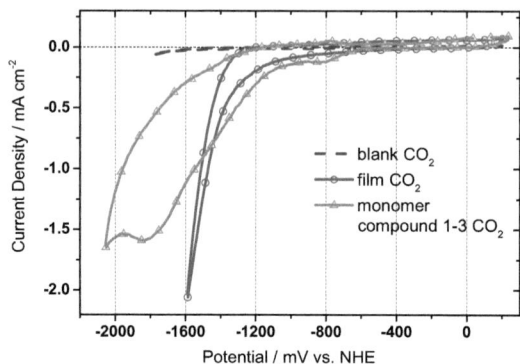

Figure 8.8: Cyclic voltammograms of the rhenium catalyst film on a Pt plate electrode (red line with circles) and a 1 mM solution of the monomer 1-3 (green line with triangles) in CO_2-saturated electrolyte solution. The scan in the presence of CO_2 shows large current enhancement for both systems due to a catalytic reduction of CO_2 to CO. Voltammograms were recorded at $100\,\mathrm{mVs^{-1}}$ in acetonitrile and a Pt counter electrode. A scan with no catalyst film present under CO_2 (blue dashed line) shows little to no reductive current.

catalyst film, headspace gas samples were taken and analyzed with regard to the CO-concentration using GC and FTIR measurements. Knowing the partial pressure of the CO formed, the number of molecules of CO dissolved in the electrolyte solution was estimated using Henry's Law following Equation 2.17. The Faradaic efficiency (η_F) was then calculated according to Equation 2.16.

With this approach, a Faradaic efficiency for the reduction of CO_2 to CO by the rhenium catalyst film of about 33% was calculated. The Faradaic efficiency of a 1 mM solution of the monomer catalyst 1-3 was measured to be around 43%. The control experiment with a pure platinum plate electrode under otherwise identical conditions did not yield detectable amounts of CO. Similar to the homogeneous catalysis it is likely that under these conditions (ACN:TBAPF$_6$) also small amounts of formate and oxalate can be formed, however with typical Faradaic efficiencies below 1%.[45]

Further characterization to determine turn over number (TON) and turn over frequency (TOF) is important. These parameters will help in determining stability and lifetime of the novel catalyst film and should be further investigated. Typical TON for Re(bipy)(CO)$_3$X (X=Cl, Br) compounds are in the order of 300.[88, 115, 46] To determine the TON of the catalyst film, specific information on the catalytic active rhenium sites on the film would be necessary. This data however is not available at the current stage of investigation and is very difficult to obtain experimentally. As a fist approximation one can assume a single homogeneous active monolayer of the catalyst film on the electrode surface. This would result in a surface coverage of about $1.5x10^{-10}$ mol active rhenium-sites per cm^2 and is in the order of similar reported catalyst films by B.P. Sullivan and T.J. Meyer et al. in 1989.[83] Dividing now the amount of CO formed during the electrolysis experiment over 60 min by the number of estimated active rhenium sites would result in about 1400 turn overs per active site and a frequency of about 0.4 turn overs per second. These values seem to be reasonable, since it is known from literature that electro-polymerized rhenium catalysts execute about 30 times more turnovers per site than their monomer counterparts in solution.

Figure 8.9(b) shows a comparison of the absorption spectra of a dilute solution of the monomer 1-3 in acetonitrile and of the rhenium catalyst film on a platinum plate electrode. In acetonitrile, the spectrum of the monomer compound 1-3 is dominated by an intense absorption maximum at around 375 nm resulting from a strong intraligand band with additional metal to ligand charge transfer contributions. Other weaker UV-bands of intraligand origin occur between 280 nm and 320 nm. A detailed study on the nature of the electronic transitions and the photophysical behaviour of the monomer compound was published by K. Oppelt et al.[62]

After electropolymerization, the rhenium catalyst film shows a red shift in the absorption band with a maximum at around 425 nm. Compared to 1-3, this red-shift of the absorption maximum in the film of the rhenium catalyst can probably be attributed to an increased delocalization of the conjugated π-electron system upon polymerization. Though, the red shift is not as significant as observed in other conjugated polymers. Therefore the effective conjugation is probably limited to a few monomer units.

The polymer growth process was also investigated with an ex-situ ATR-FTIR technique. For this measurement, a ZnSe reflection element covered with a thin (10 nm) sputtered film of platinum was used as working electrode, and a 150 nm layer of the rhenium catalyst film was potentiostatically electropoly-

8.2. (5,5'-BISPHEN.-2,2'-BIPY.)RE(CO)$_3$CL POLYMERIZATION

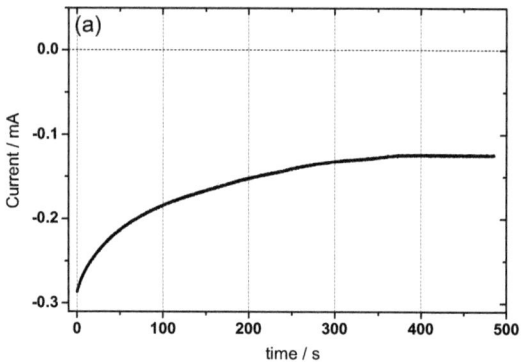

(a) Current time curve

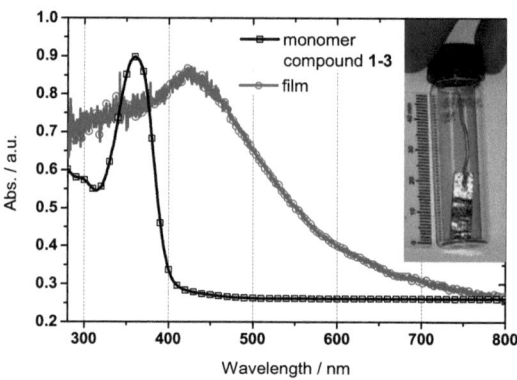

(b) UV-Vis absorption spectra

Figure 8.9: (a) Current time curve for the potentiostatic film formation in nitrogen saturated acetonitrile solution containing 0.1 M TBAPF$_6$ initial monomer1-3 concentration of 2 mM at a constant potential of -1550 mVvs.NHE. (b) Comparison of the UV-Vis absorption spectra of the rhenium catalyst film on a platinum plate electrode (red solid line) and of a $6.25 \cdot 10^{-5}$ M solution of the monomer 1-3 in acetonitrile (298 K, 1 cm cell, black dots) (left side). On the right side a picture of the rhenium catalyst film electropolymerized onto a Pt-plate electrode is shown.

merized on the Pt-surface of the modified ZnSe ATR crystal. The experiment was performed in a one compartment electrochemical cell as depicted in the *Experimental section* in Scheme 2.17(A). The electrolyte solution contained 0.1 M TBAPF$_6$ initial monomer1-3 concentration of 2 mM in acetonitrile. The electrochemical cell was connected to the potentiostat and a constant potential of -1550 mV vs. NHE was applied for 500 seconds. The electrochemical current measured as function of time is presented in Figure 8.9(a). As can be seen, during film formation, the current dropped from initially -0.3 mA to approx. -0.12 mA after 400 sec. and stayed constant afterwards. This indicates that the film formation ceased at that time.

After electropolymerization, the ZnSe/Pt electrode with the 150 nm thick catalyst film was mounted into an ATR-FTIR setup between two PTFE spacers (see in the *Experimental section* in Scheme 2.17(B)) and the ATR-FTIR difference absorption spectra between a pure ZnSe/Pt electrode and the ZnSe/Pt electrode with the catalyst film were recorded.

Figure 8.10 shows the ATR-FTIR difference absorption spectrum of the 150 nm thick film of the rhenium catalyst vs. a pure ZnSe/Pt electrode and of the monomer 1-3 dissolved in dichloromethane and drop cast on a ZnSe ATR crystal vs. the pure ZnSe ATR crystal. In the spectrum of the rhenium catalyst film distinctive new peaks are observed and their peak positions are indicated by the numbers 1 to 10 in the absorption spectrum of Figure 8.10. These positions are connected with characteristic vibrations of the polymerized film. As compared to the infrared spectrum of the monomer (bottom, black solid line) there is a noticeable decrease in the intensity of the monomer main peaks 3 and 4 positioned at around 1900 cm^{-1} and 2000 cm^{-1} which are characteristic signals connected to the C≡O vibrations.[62] The number of peaks and their relative intensities, however, do not change, which indicates that the incorporated catalytic centers of the monomer 1-3 are not decomposed upon polymerization, and that a facial arrangement (fac-isomeric form) of the carbonyl ligands is obviously retained at the rhenium atom.

In contrast, the IR-signal positioned at 2200 cm^{-1} that is characteristic for the C≡O vibrations almost vanishes completely, while the appearance of a new sharp peak 8, centered at around 848 cm^{-1}, indicates the asymmetric stretching band of PF$_6^-$. It is assumed that the vanishing of the peak at 2200 cm^{-1} might be directly connected to a loss of the C≡O signal in the course of the polymerization process. The fact that no solvent bands from acetonitrile at 2293, 2252, 1442, 1035 and 917 cm^{-1},[128] are present in the spectrum suggests that the 848 cm^{-1} band can be attributed to the presence of PF_6^- anion containing species in the polymer phase. The strong increase of this peak could then be explained by a substitution of the Cl^- in 1-3 by PF_6^- from the electrolyte during polymerization to retain charge neutality,[129] or because of the rather high relative intensity of the signal, by an uptake of a certain amount of the alkylammonium electrolyte. The two peaks 1 and 2, at 2972 and 2877 cm^{-1}, are attributed to the valence vibrations of additionally aliphatic C-H bonds formed either during polymerization, or due to the presence of the alkylammonium salt. The peak 5, at 1640 cm^{-1}, is characteristic for aryl-conjugated C=C bonds. The two smaller peaks 9 and 10, at 750 and 690 cm^{-1}, are typical for out-of-plane C-H bending vibrations of monosubstituted benzene.[130] Additionally the peak 7 at 1065 cm^{-1} might originate from not well defined C-H vibrations along the main chain of the polymer film, which would explain the relative broadness of this absorption peak.

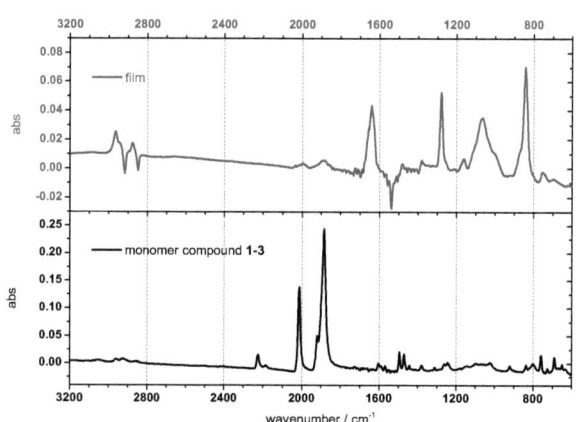

Figure 8.10: ATR-FTIR difference absorption spectra of a 150 nm thick rhenium catalyst film on 10 nm Pt sputtered onto a ZnSe ATR crystal to the pure 10 nm Pt/ZnSe ATR crystal (top, red line) and of the monomer 1-3 dissolved in DCM and drop cast on a ZnSe ATR crystal to the pure ZnSe ATR crystal (bottom black line).

Following this argumentation, it is assumed that the polymerization proceeds via radical addition similar to the mechanism published by Canadas et al.[130] Still, due to the strong negative potential necessary for electropolymerization of the monomer 1-3, it is likely that additional side reactions take place that account for the unassigned peak 6 at $1280\,cm^{-1}$ in the IR-spectrum of Figure 8.10. As a result, at this time it is not yet possible to determine the exact structure of the rhenium catalyst film. For the catalytically active centres of the rhenium catalyst film however, we propose a structural motif similar to the one depicted in Scheme 8.4.

The film morphology was studied using SEM and AFM technique. For this purpose the film surface was scratched to see the difference between the rhenium catalyst film formed (and the ZnSe/Pt substrate. As was reported in a published paper,[131] a highly ordered granular structure is observed and attributed to the granular structure is characteristic for sputtered platinum. However, the type of structure observed for the catalyst film is significantly different from the metallic substrate. The lack of clearly visible structures on the surface of the film suggests a lack of order in the film, which is expected for polymeric films. From a measurement of the tilted substrate by an angle of $54.0°$, the film thickness of $150\,nm$ could be measured. The roughness of the film was studied by atomic force microscopy (AFM). A close look on the film surface reveals the existence of particles that probably remain from the film formation process. The RMS roughness in the regions without particles was found to be $5.4\,nm$.

Concluding the electropolymerisation of the monomer 1-3 it appears to be a promising way to change from homogeneous to heterogeneous catalysis for CO_2-reduction, reducing the amount of required catalyst material needed and overcoming the limitations regarding solubility of homogeneous catalysts in general. The novel catalyst film furthermore demonstrates high selectivity for the CO_2-reduction to CO at relatively low reduction potentials and high current densities. A detailed study on the polymerization of 5,5'-bisphenylethynyl-2,2'-bipyridyl)Re(CO)$_3$Cl and its characterization has been published.[131]

8.3 (4,4'-dicarboxyl-2,2'-bipy.)Re(CO)₃Cl immobilization at a zinc oxide layer

A different approach was used at the very beginning of the thesis when the rhenium catalyst 1-2 with the dicaboxylic side group at the ligand system was used to attach to a binding metal oxide layer with the aim of moving from a homogeneous to a heterogeneous catalysis (compare Figure 6.8). Zn was chosen as working electrode due to its high overpotential for hydrogen evolution and its native oxide layer. Scheme 8.11 illustrates the idea of anchoring the rhenium catalyst 1-2 onto an ZnO layer with the purpose of going from homogeneous to heterogeneous catalysis. Additionally in Scheme 8.11 the series from a to c depicts the expected, subsequent CO_2 reduction mechanism upon negative bias of the electrode in aqueous solution saturated with CO_2. The mechanism was expected to proceed similar to the homogenous system, namely a two electron, proton coupled CO_2 reduction according to reaction 1.4 described in the introduction chapter 1, leading to CO and H_2O as reduction products.

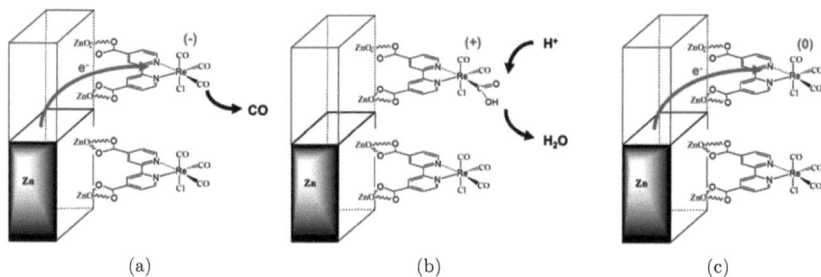

Figure 8.11: Schematic representation of (4,4'-dicarboxyl-2,2'-bipyridyl)Re(CO)₃Cl (1-2) attached to an electrode by the metal native ZnO layer and the expected, subsequent CO_2 reduction mechanism a to c.

Figure 8.12 shows the cyclic voltammograms of (4,4'-dicarboxyl-2,2'-bipyridyl)Re(CO)₃Cl immobilized at a zinc oxide layer in aqueous, N_2 purged solution with a $KHPO_4$ buffer at pH 7. Below a potential of approx. -1200 mV vs. Ag/AgCl in 3M KCl a reductive wave is observed attributed to the reduction of the metaloxide-catalyst bond. Below a potential of -1400 mV vs. Ag/AgCl substantial reductive current is observed due to hydrogen (H_2) evolution.

It can be concluded, that in this experiments the anchoring to the oxide layer

CHAPTER 8. HETEROGENEOUS ELECTRO CATALYSIS

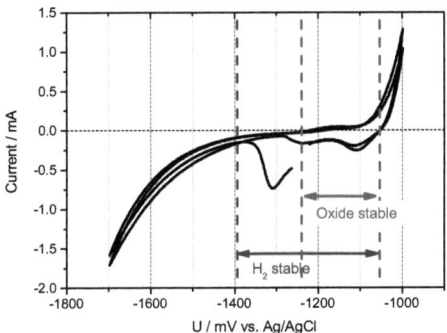

Figure 8.12: Cyclic voltammograms of (4,4'-dicarboxyl-2,2'-bipyridyl)Re(CO)$_3$Cl immobilized at a zinc oxide layer in aqueous, N$_2$ purged solution with a KHPO$_4$ buffer at pH 7

was successfully achieved, however it tourned out that the film was not stabel uppon electrochemical reduction. The cyclicvoltammetry experients showed, that below a potential of -1200 mV vs. Ag/AgCl or equivalent -1000 mV vs. NHE (compare Figure 8.12) the layer was irreversibly destroyed and rinsed of the electrode as a black precipitate. Similar experiments have been reported where the anchoring to the oxide layer of TiO$_2$ was successfully achieved [127] showing the working principle of the idea, however up to now no application for CO$_2$ reduction was reported. This work has not been carried further within the frame of this thesis.

8.3. (4,4'-DICARBOXYL-2,2'-BIPY.)RE(CO)$_3$CL IMMOBILIZATION

Chapter 9

Organic semiconductors for CO_2 reduction

As described in the introduction chapter under 1.4, a very promising approach in the last decade was the combination of catalysts with inorganic semiconductor materials. In this work the approach is different in a way that an organic semiconductor is used as light absorbing donor material in combination with an acceptor catalyst, as for example Pyridinium or (2,2'-bipyridyl)Re(CO)$_3$Cl, for the actual CO_2 reduction. The advantage lies in the unique properties of organic semiconductors as they are flexible, light weight, bio-degradable and bio-compatible, abundant and hold hence the promising perspective to be cheap in production. Figure 9.1 shows the general idea of a photoinduced charge transfer from a biased organic semiconductor onto a catalyst redox mediator for CO_2 reduction.

The schematic drawing in Figure 1.5 shows an organic p-type semiconductor, as for example poly(3-hexylthiophene) (P3HT) on indium tin oxide (ITO), which serves as a transparent conducting electrode (TCE). The whole system is in contact with an electrolyte solution forming a Schottky type of contact.[34] The electrolyte solution contains a catalyst acceptor molecule as for example pyridinium. Upon light irradiation an electron-hole pair is generated in the organic semiconductor. The electron-hole pair generated can not be treated as individual charge carriers since the positive hole (h^+) and the negative electron (e^-) are initially still bound in the material by their Coulomb attraction.[35] The electron-hole pair, also known as exciton, moves as a quasi particle in the bulk of the material until the two charge carriers recombine or hit an interface

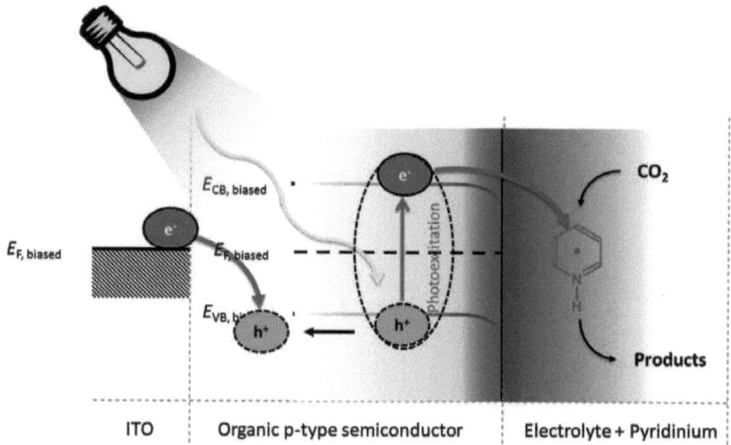

Figure 9.1: Suggested mechanism for photoinduced charge transfer from a biased organic p-type semiconductor onto pyridinium as a catalyst redox mediator for CO_2 reduction.

where the driving force is strong enough for both charges to separate. In the ideal case the exciton will reach the semiconductor-electrolyte interface where the electron transfers to the catalyst acceptor molecule and the hole, now free to move in the bulk material, travels back to the biased ITO contact where it recombines with an electron. The catalyst material loaded with the negative charge is then capable of transferring this charge to the substrate, namely CO_2, and reducing it to the desired product.

9.1 Semiconductor – electrolyte interface

The most important difference between semiconductors and metals lies in the band occupancy. Unlike metals, the electron bands in semiconductors are, at least at 0 K, either completely occupied or completely empty. For a schematic representation of the band structure of a semiconductor and the corresponding electron distribution at temperatures above 0 K compare Figure 9.2(a). For an undoped semiconductor the Fermi level (E_F) lie close to the mid-point between valence and conduction band. For a p-type semiconductor with unoccupied acceptor levels in the band gap, the Fermi level shifts towards the valence band.

CHAPTER 9. ORGANIC SEMICONDUCTORS FOR CO_2 REDUCTION

For an n-type semiconductor with occupied donor levels in the band gap, the Fermi level shifts towards the conduction band respectively, compare Figure 9.2(b).

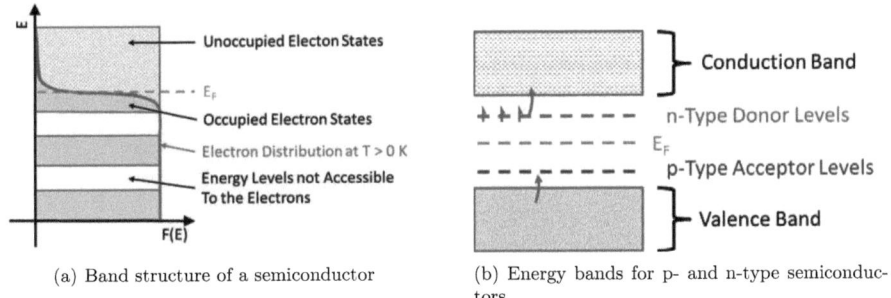

(a) Band structure of a semiconductor

(b) Energy bands for p- and n-type semiconductors

Figure 9.2: (a) Schematic representation of the band structure of a semiconductor and the corresponding electron distribution at temperatures above 0 K. (b) Schematic representation of energy bands for p- and n-type semiconductors with p-type acceptor and n-type donor energy levels respectively.

A semiconductor in the contact with an electrolyte solution forms a kind of *Schottky type of barrier* similar to a metal-semiconductor interface. When a p-type semiconductor is brought into contact with two metals of different workfunctions (eg. Ca/Au) or a metal and the electrolyte containing a redox couple, the field distribution will be initially inhomogeneous due to mobile charges in the doped semiconductor. These charges can move in the electric field. To reach equilibrium, charges will move across the metal/semiconductor/electrolyte interfaces and the majority carriers (i.e. positive charge carriers for a p-type semiconductor) are depleted from the semiconductor in the region near the interface. A space charge region is formed. The electric field is compensated. At the metal-semiconductor interface a potential drop of HOMO and LUMO is observed. The distance of this drop from the interface potential to the semiconductor flat band potential (φ_{fb}) is called the depletion width (W) , which will increase under reverse bias. Under forward bias, the depletion with is reduced, eventually to nothing and carriers can flow freely through the device, compare Figure 9.3. [34, 132, 133]

In other words, when the semiconductor is brought into contact with the electrolyte solution, the energy of the electrons in the highest occupied levels of the band must have the same free energy as the redox couple in solution.

9.1. SEMICONDUCTOR – ELECTROLYTE INTERFACE

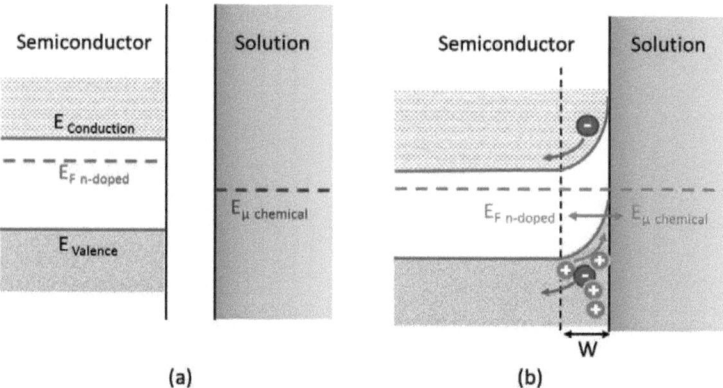

Figure 9.3: Band structure for a n-type semiconductor not in contact with the electrolyte solution (a) and (b) band-bending if the semiconductor is in contact with the electrolyte.

This leads to the band bending of the semiconductor. The same applies for a metal-semiconductor interface where the chemical potenials of metal and semiconductor are $\mu_{e-metal} = \mu_{e-Semiconductor}$. A Schottky type of barrier is formed. According to Schottky, the rectification of a metal-semiconductor device is dependent from the space charge in the semiconductor. This means, that the characteristic of the metal-semiconductor junction corresponds to the doping concentration N in the semiconductor.

The total space charge in the semiconductor is then given by equation 9.1.

$$Q_{SC} = q \cdot N_A \cdot W \tag{9.1}$$

And hence the capacitance in the space charge region can be expressed by equation 9.2 and equation 9.3.

$$C_{SC} = \frac{dQ_{SC}}{dV} \tag{9.2}$$

or:

CHAPTER 9. ORGANIC SEMICONDUCTORS FOR CO_2 REDUCTION

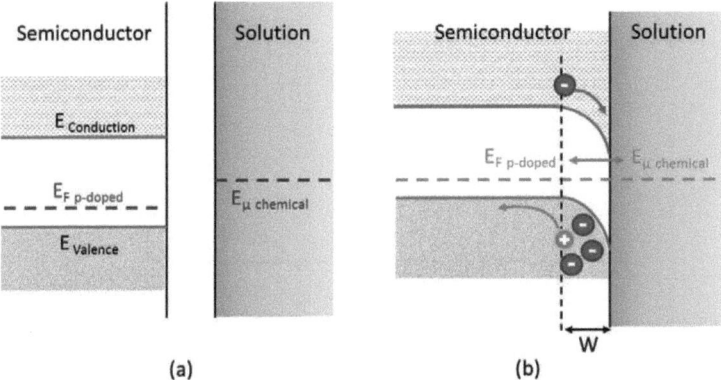

Figure 9.4: Band structure for a p-type semiconductor not in contact with the electrolyte solution (a) and (b) band-bending if the semiconductor is in contact with the electrolyte.

$$\frac{1}{C_{SC}^2} = \frac{2\left(V_{applied} - \varphi_{fb} - \frac{kT}{q}\right)}{q\epsilon_0\epsilon_R N_A} \qquad (9.3)$$

Where $V_{applied}$ is the applied potential, φ_{fb} is the flat-band potential and the other symbols have their usual meaning. Therefore, if the carrier concentration (N_A) is constant, a plot of $1/C^2$ vs. voltage (from impedance measurements) is expected to show a linear regime, compare Figure 9.5. This is known as the Mott-Schottky analysis that reveals the flat-band potential φ_{fb} where the line intersects the potential axis and the slope k can be used to obtain the doping level (donor density).[134, p. 136-139]

As for metallic electrodes, changing the potential applied to the electrode shifts the Fermi level accordingly. The band edges in the interior of the semiconductor, i.e. away from the depletion region, also vary with the applied potential in the same way as the Fermi energy. However, the energies of the band edges at the interface do not change by changes in the applied potential.[133] For a detailed analysis of semiconductor interface physics see ref. nr. [134].

Similar for organic semiconductors, if an external bias is applied to the polymer one moves the chemical potential μ_{e-} up or down from the ideal, unperturbed system. For a semiconductor the chemical potential can be anywhere in the gap. If a positive potential is applied, the chemical potential will move

9.1. SEMICONDUCTOR – ELECTROLYTE INTERFACE

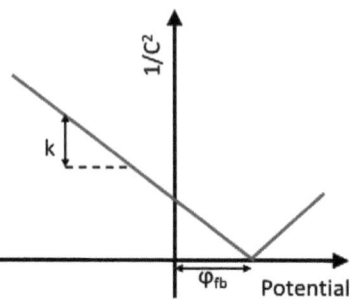

Figure 9.5: Schematic representation of the *Mott – Schottky* analysis for a p-type semiconductor.

down, i.e. closer to the valence band. When the chemical potential reaches the π-valence band of the organic semiconductor, the material gets oxidized, i.e. charges are pulled out of the material. If the applied voltage is negative, vice versa μ_{e-} shifts upwards to the conduction band and reduction occurs once μ_{e-} reaches the π^*-conduction band of the organic semiconductor. However, typically in organic semiconductors there are states in the gap that pin the Fermi-level and it is difficult to move the chemical potential by an externally applied bias. In experiments one finds that the energy gap between electrochemical oxidation and reduction is very close to what one finds in optical absorption measurements, compare Figure 9.6.[35]

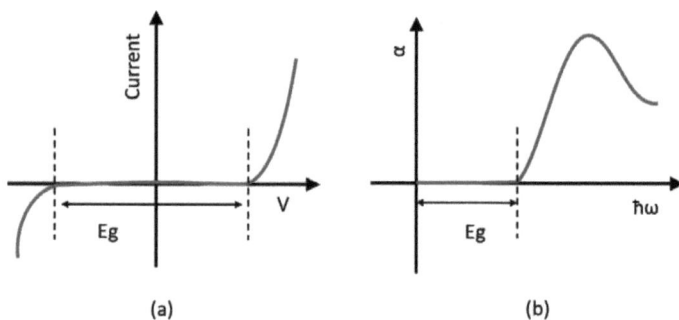

Figure 9.6: Schematic representation of (a) electrochemical reduction and oxidation and (b) optical absorption measurement.

Such experimental analysis for a real system, namely compound (2,2'-bipyridyl)Re(CO)$_3$Cl (1-1) are shown in Figure 6.4 (electrochemical oxidation

CHAPTER 9. ORGANIC SEMICONDUCTORS FOR CO_2 REDUCTION

and reduction) and Figure 6.5(b) (optical absorption measurement) in section 6.

9.2 Poly(3-hexylthiophene) electrodes with pyridinium as homogeneous catalyst

In this section poly(3-hexylthiophene) (P3HT) is used as the organic semiconducting polymer which serves as donor material and Pyridinium, already described in chapter 6, as a homogeneous electrocatalyst in solution. Figure 9.7(a) shows the schematic chemical structure of poly(3-hexylthiophene) (P3HT). Initially it is important to determine the reductive potential window of such a system in order to know up to what negative potential the working electrode can be reversibly biased without damage of the electrode or the electrolyte and is usually defined by the potential range (window) were no significant current is observed.

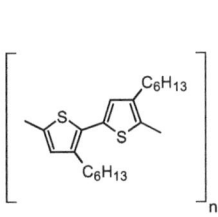

(a) Poly(3-hexylthiophene) (P3HT)

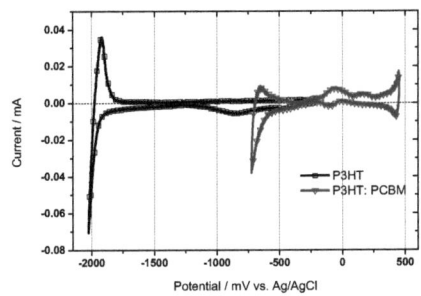

(b) Cyclic voltammograms of P3HT on ITO

Figure 9.7: (a) Schematic chemical structure of poly(3-hexylthiophene) (P3HT) (b) Cyclic voltammograms of a P3HT (black line with squares) and a blend of P3HT:PCBM (blue line with triangles) covered ITO working electrode. Experiments with a blended P3HT:PCBM working electrode demonstrate a substantial smaller reductive potential window. Voltammograms are recorded in a nitrogen saturated acetonitrile solution containing 0.1 M TBAPF$_6$ and a scan speed of 50 mVs^{-1}.

Figure 9.7(b) shows the cyclic voltammograms of a P3HT (black line with squares) and a blend of P3HT:PCBM (blue line with triangles, PCBM = Phenyl-C61-butyric acid methyl ester) covered ITO working electrode. It is

9.2. POLY(3-HEXYLTHIOPHENE) ELECTRODES WITH PYRIDINIUM

known that a mixture of P3HT and PCBM is beneficial in a way that the PCBM works as an efficient acceptor molecule for electrons in a P3HT:PCBM blend and hence increasing the photocurrent supported by those bulk-heterojunction layers significantly.[135] For electrochemistry and photoelectrochemistry experiments it was found however, that a blended P3HT:PCBM as working electrode on ITO demonstrate a substantial smaller reductive potential window as can be seen in Figure 9.7(b) (blue line with triangles) which makes this combination not suitable for electrochemical applications. Therefore, in the following experiments a pure P3HT film on ITO/glass substrate was used.

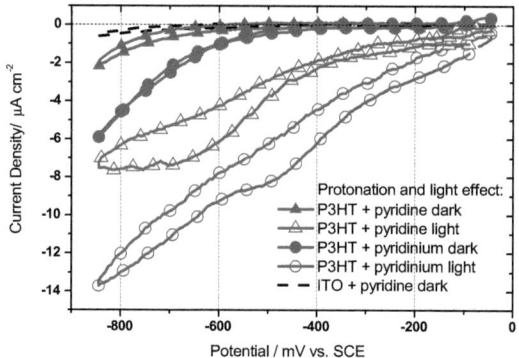

Figure 9.8: Cyclic voltammetry of a 100 mM pyridine (green lines with triangles) and pyridinium (red lines with circles) electrolyte solution under dark and white light illumination with a 120 nm thick P3HT covered ITO working electrode. Experiments under illumination (open symbols) show a significant current increase compared to the corresponding experiment in dark. For comparison a scan with a pure ITO electrode under otherwise identical conditions is also shown (black dashed line). Voltammograms are recorded at $10\,\mathrm{mVs^{-1}}$ in a nitrogen-saturated KCl solution (0.1 M) and a Pt counter electrode.

Figure 9.8 shows cyclic voltammograms of a 100 mM pyridine (green lines with triangles) and pyridinium (red lines with circles) electrolyte solution under dark and white light illumination with an approximately 120 nm thick P3HT film covered on ITO as working electrode. The experiment shows the effect upon protonation and light. It can be seen that the experiments under illumination show a significant current increase compared to the corresponding experiment in dark. Already when pyridine is present, the current increases significantly, which leads to the conclusion that pyridine acts as an acceptor molecule for

CHAPTER 9. ORGANIC SEMICONDUCTORS FOR CO_2 REDUCTION

charge transport. When the pyridine is converted to pyridinium, by the addition of small amounts of diluted sulfuric acid, the dark current increases which can be attributed to proton reduction. Upon light irradiation there is again a clear increase in the reductive current especially in the potential range where no current could be observed under dark conditions. For comparison a scan with a pure ITO electrode under otherwise identical conditions is also shown (black dashed line).

To investigate this effect further, potentiostatic experiments were performed where the semiconducting working electrode was biased to a constant potential of -800 mVvs. SCE and the light was repeatedly turned on an off.

Figure 9.9(a) shows the photoresponce of a P3HT covered ITO electrode at constant potential of -800 mVvs. SCE upon white light irradiation and a varying the pyridinium concentration from 10 to 100 mM. Upon light irradiation the current jumps immediately to more negative values and returns to the initial value when the light is turned off again. When the catalyst concentration is increased, the corresponding current photo-response is enhanced, which is expected when pyridinium acts as an efficient acceptor molecule. In Figure 9.9(b) the photocurrent of the data shown in Figure 9.9(a) is plotted as $\Delta j_{light-dark}$. It can be seen that the photocurrent increases linearly with the pyridinium concentration between 10 mM and 100 mM (black dashed line).

In Figure 9.10 a similar experiment is shown where the photocurrent (as $\Delta j_{light-dark}$) from the photoresponse experiments of a P3HT covered ITO electrode at constant potential of -800 mVvs. SCE upon white light irradiation is plotted at different light intensities. The experiment was performed for two different pyridinium concentrations, namely a 10 mM and a 100 mM concentration. It was found that the photocurrent increases with increasing light intensity for both concentrations of the catalyst acceptor molecule.

One potential error of importance is the temperature effect upon light irradiation. As the solvent and electrode absorb some portion of the white light irradiation it can not be excluded that a temperature effect is observed. By taking a look into the theory of electrochemistry it appears to be very difficult describing mathematically the dependence of the diffusion limited current density of an electrochemical cell on the electrode temperature. This is mainly due to the fact that the kinetics of the overall electrochemical reaction are not only

9.2. POLY(3-HEXYLTHIOPHENE) ELECTRODES WITH PYRIDINIUM

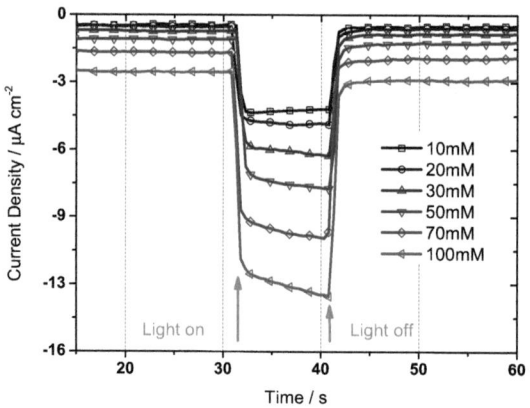

(a) Photoresponce of P3HT:Pyridinium

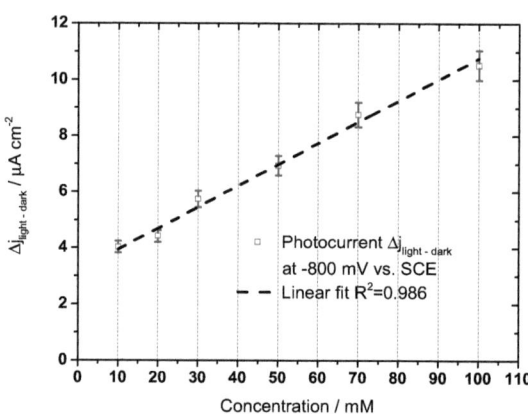

(b) Photocurrent plotted as $\Delta j_{light-dark}$

Figure 9.9: (a) Photoresponce of a P3HT covered ITO electrode at constant potential of -800 mVvs. SCE upon white light irradiation and a varying pyridinium concentration from 10 to 100 mM. The measurements are recorded in a nitrogen-saturated KCl solution (0.1 M) and a Pt counter electrode. (b) Photocurrent plotted as $\Delta j_{light-dark}$ from the photoresponse experiments depicted in (a). The photocurrent increases linearly with the pyridinium concentration between 10 mM and 100 mM (black dashed line).

CHAPTER 9. ORGANIC SEMICONDUCTORS FOR CO_2 REDUCTION

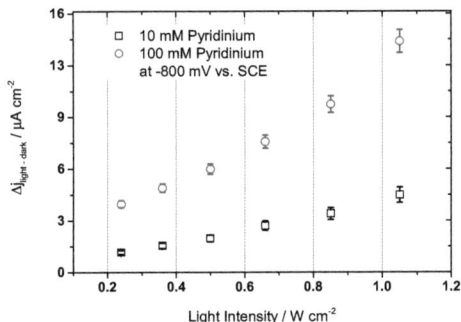

Figure 9.10: Photocurrent plotted as $\Delta j_{light-dark}$ from the photoresponse experiments of a P3HT covered ITO electrode at constant potential of -800 mVvs. SCE upon white light irradiation at different light intensities. The photocurrent increases with increasing light intensity for a pyridinium concentration of 10 mM (black squares) and 100 mM (red circles).

determined by the electron transfer process, but also by transport processes like diffusion, ad- and desorption of reactants and reactions in the homogeneous solution preceding or following the charge transfer. A very good description of the complexity of such a system is described in reference nr. [136]. To estimate this error an additional experiment was performed where a thermocouple was attached to a blank ITO electrode in an otherwise identical system. It is expected that the ITO serves as a good heat conductor giving an almost homogenoeous heat distribution within the electrode area of approximately 4 cm².

Figure 9.11 shows the temperature increase upon white light irradiation on a ITO electrode in a nitrogen-saturated KCl solution (0.1 M) (blue triangles). As can be seen the temperature increase follows an exponentially increasing behavior according to a fitting curve of
$y(t) = 26.66 \cdot (1 - e^{-0.0034 \cdot (t+296.7)})$ (black line). It can be seen that within an illumination time of 20 min. the temperature increased by about 6.6 °C. Analyzing this behavior the electrode temperature has a time constant (τ) of about 294.12 s which represents the time it takes the system's response to reach $1 - 1/e$ (about 63.2 %) of its final (asymptotic) value.

In a very detailed study on the temperature effect of the working electrode on the limiting current of a 2 mM solution of $Ru(NH_3)_6^{3+}$ in aqueous 0.1 M KCl

9.2. POLY(3-HEXYLTHIOPHENE) ELECTRODES WITH PYRIDINIUM

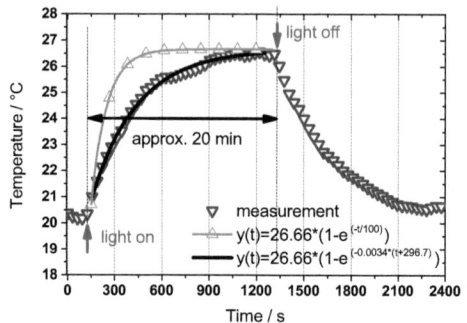

(a) Temperature increase measurement

(b) Experiment foto

Figure 9.11: (a) Temperature increase upon white light irradiation on a ITO electrode in a nitrogen-saturated KCl solution (0.1 M) (blue triangles). The temperature increase follows $y(t) = 26.66 \cdot (1 - e^{-0.0034 \cdot (t+296.7)})$ (black line). (b) Foto of the corresponding experimental setup.

it was shown that upon a temperature increase of 6.7°C from 303.9 K to 310.6 K (about the same temperature range as in the experiment in Figure 9.8) the relative increase in the limiting current density was about 16.8 %.[137] Since in the cyclic voltammetry experiments depicted in Figure 9.8, the scan speed was $10\,\mathrm{mVs^{-1}}$ and the return potential is - 850 mV, reaching peak current takes 85 s. At this return potential the relative increase in current density between dark and light measurement is about 133 %. Following this argumentations it can be safely concluded that a temperature effect can not be responsible for the change in the cyclic volltammetry current densities upon light irradiation depicted in Figure 9.8, nor for the immediate photoresponse observed in the measurements shown in Figure 9.9 and Figure 9.10.

Bulk electrolysis experiments using the P3HT/pyridinium system were performed, however no products in form of methanol could be detected. This is not surprising since the lifetime of this type of electrodes is rather low and the photocurrent is in the μA regime. If one has a look back to chapter 6 where the pyridinium catalyst was described, one finds for example in Figure 6.22(a) that the current densities were in the order of mA for electrolysis experiments of 30 hrs and only then it was possible to detect methanol in the electrolyte solution with concentrations in the ppm level. This concentrations are close to the detection limit of the GC system. In the case with the P3HT covered ITO electrodes the electrodes corroded within 2 to 3 hrs of electrolysis time. As a resulting conclusion, this system has to be improved significantly in terms of electrode stability and current densities which is subject to ongoing research.

9.3 Polyvinylcarbazole electrodes with (2,2'-bipy.)Re(CO)$_3$Cl as homogeneous catalyst

It has been mentioned in chapter 7, *Homogeneous photo catalysis*, that the fac-(2,2'-bipyridyl) Re(CO)$_3$Cl (1-1) is an efficient photo catalyst for the reduction of CO$_2$ producing mainly CO with the aid of a sacrificial electron donor as for example triethanolamine (TEOA). The quantum yield (Φ_{CO}) for CO$_2$ reduction of this compound 1-1 has been reported with 0.14 almost 30 years ago and subsequent modifications of the bipyridine ligand system of 1-1 lead to the development of superior catalysts with Φ_{CO} up to 0.59, making these type of

9.3. POLYVINYLCARBAZOLE WITH (2,2'-BIPY.)RE(CO)₃CL

materials the most efficient CO_2 photo-catalyst among known homogeneous catalyst materials by now.[47, 58, 88, 111]

Unfortunately rhenium tricarbonyl complexes have several disadvantages when it comes to practical applications, stating their poor absorption in the visible region of the solar spectra and the low abundance of rhenium as the most prominent ones. One promising solution to overcome this disadvantages of rhenium based catalysts is to use poly(N-vinylcarbazole) (PVK) as a decent an cheap absorber to serve as redox photosensitizer in combination with fac-(2,2'-bipyridyl)Re(CO)₃Cl (1-1) as catalyst acceptor.

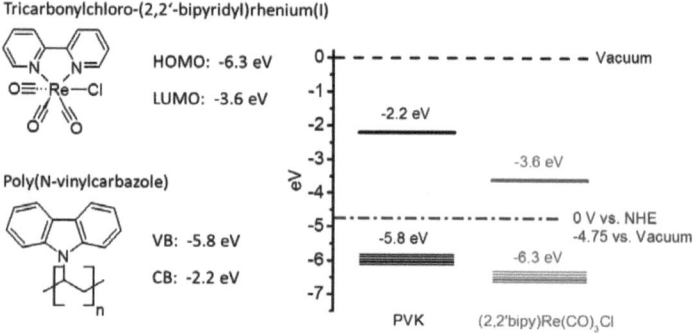

Figure 9.12: (a) Chemical structure of (2,2'-bipyridyl)Re(CO)₃Cl and its corresponding highest occupied molecular orbital (HOMO) and lowest unoccupied molecular orbital (LUMO) level in electron volt (eV), and Poly(N-vinylcarbazole) (PVK) and its corresponding energy values for valence (VB) and conduction band (CB). (b) Schematic drawing of frontier orbital energy levels comparing PVK as donor and (2,2'-bipyridyl)Re(CO)₃C (1-1) as acceptor molecule.

Figure 9.17(a) shows the chemical structure of (2,2'-bipyridyl)Re(CO)₃Cl and its corresponding highest occupied molecular orbital (HOMO) and lowest unoccupied molecular orbital (LUMO) level determined with -6.3 eV vs. vacuum for the HOMO and -3.6 eV vs. vacuum for the LUMO respectively. The values are determined from electrochemical and UV-Vis absorption measurements as well as quantum mechanical calculations on the DFT-level compare chapter 4 and chapter 6 of the thesis. Additionally Figure 9.17(a) shows the chemical structure of poly(N-vinylcarbazole) (PVK) and its corresponding energy values for valence (VB) and conduction band (CB) with -5.8 eV vs. vacuum and -2.2 eV vs. vacuum respectively, taken from literature values.[138, 139]

CHAPTER 9. ORGANIC SEMICONDUCTORS FOR CO_2 REDUCTION

Figure 9.17(b) shows the schematic drawing of the material frontier orbital energy levels comparing PVK as donor and (2,2'-bipyridyl)Re(CO)$_3$Cl as acceptor molecule versus the vacuum energy. For the electrochemical consideration of frontier orbital energy levels an offset of the normal hydrogen electrode (NHE) potential vs. the vacuum leve with -4.75 eV is used (blue dashed line).[42] As can be seen in the schematic of Figure 9.17(b) the energy levels of the donor polymer and acceptor catalyst are aligned in a favorable situation for photo-excited charge and/or energy transfer. Subsequent photoluminescence (PL) quenching and light induced ESR experiments were carried out to study this system in bulk, solid phase mixtures of donor and acceptor, and at a donor-acceptor solid-liquid interface.

9.3.1 Photoluminescence quenching in solid films

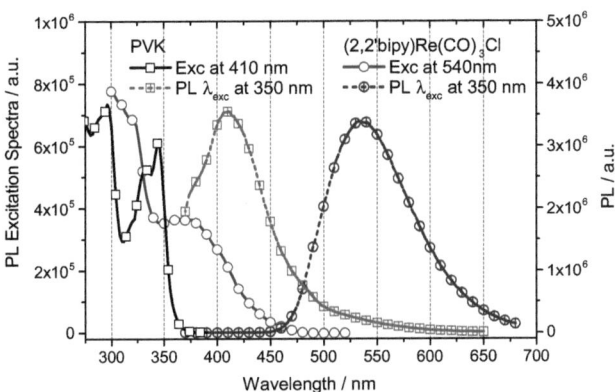

Figure 9.13: PVK and (2,2'-bipyridyl)Re(CO)$_3$Cl (1-1) excitation and photoluminescence spectra. Excitation spectra for a pure, approx. 60 nm thick film of PVK (black solid line with squares), photoluminescence of the same PVK film on glass substrate excited at 350 nm (dashed gray line with squares), excitation spectra for a pure, approx. 40 nm thick film of 1-1 (blue line with circles) and photoluminescence of the same film of 1-1 on glass substrate excited at 350 nm (dashed dark blue line with circles).

Figure 9.13 shows the excitation and photoluminescence spectra of PVK and catalyst 1-1. The excitation spectra for a pure, approx. 60 nm thick film of PVK (black solid line with squares) shows the detector signal at 410 nm of the photoluminescence while the excitation wavelength is scanned from 250 to 380 nm. The spectra shows clear separated maxima at approximately 340 and 300 nm with a distinct shoulder at 330 nm. The dashed gray line with squares give the corresponding photoluminescence of the same PVK film on glass substrate excited at 350 nm with a maximum at 410 nm. Additionally Figure 9.13 shows the excitation spectra for a pure, approx. 40 nm thick film of (2,2' bipy)Re(CO)$_3$Cl (blue line with circles) which gives the detector signal at 540 nm of the photoluminescence while the excitation wavelength is scanned from 300 to 520 nm. The spectra shows the typical strong electronic $^1\pi\pi^*$ intraligand transitions of the diimine ligands in the higher energetic region, at wavelengths shorter than 330 nm and MLCT signatures at lower energies with a maximum at about 370 nm.[140, 62] The dashed dark blue line with circles give the photoluminescence of the same (2,2' bipy)Re(CO)$_3$Cl film on glass substrate excited at 350 nm. The broad emission of the ^{3}MLCT to the ground state of the compound 1-1 is covering a typical wide spectral region including blue, green and yellow light with its maximum at around 540 nm. The excitation spectra are normalized to the spectra of the excitation lamp (photons per wavelength) measured by an internal, calibrated silicon diode.

Figure 9.14(a) shows PL quenching experiments with PVK as donor polymer and (2,2' bipy)Re(CO)$_3$Cl catalyst as quenching material on glass/ITO substrate. The materials were dropcast from dichloromethane (DCM) solution. The PL spectra are corrected by the absorption of each film taking an average value of 10% for the contribution due to light scattering into account. The amount of PVK is in all samples $5 \cdot 10^{-4}$ g. In Figure 9.14(a), the solid line with black squares give the photoluminescence of a pure, approx. 1.8 μm thick, PVK film on glass substrate excited at 350 nm. It shows a clear PL maximum at about 410 nm. As soon as some catalyst material is present the photoluminescence of the PVK polymer is reduced and the typical catalyst photoluminescence with its maximum at 553 nm appears. At a quencher concentration of approx. $15 \cdot 10^{-6}$ g (about 3% of the total material in the film) the PL signal of the PVK donor polymer disappeared completely and only the PL signal of the acceptor catalyst at 550 nm is present. In Figure 9.14(b) the PL signal of the data in (a) is subtracted by the PL of the pure PVK. This plot reveals a clear isosbestic point at 485 nm as indicated in the graph. This indicates that the change in the PL signal scales equally to lower and higher wavelengths of the isosbestic point which also suggests that there is little to no degradation during the measurement.

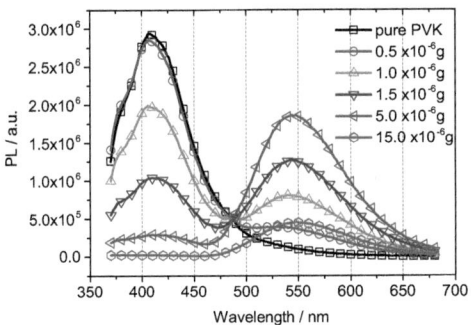

(a) PL quenching experiments with PVK

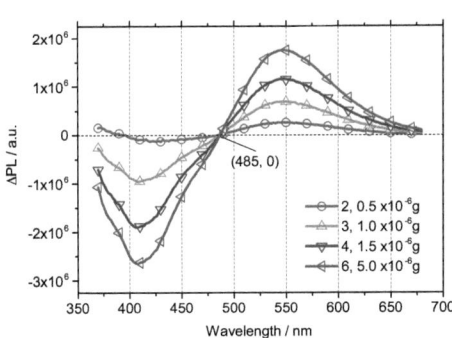

(b) ΔPL of data in (a)

9.3. POLYVINYLCARBAZOLE WITH (2,2'-BIPY.)RE(CO)$_3$CL

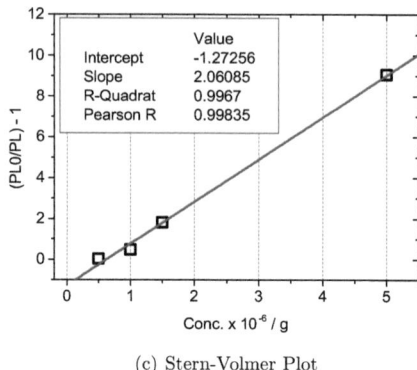

(c) Stern-Volmer Plot

Figure 9.14: (a) PL quenching experiments with PVK as donor polymer and different concentrations of (2,2'-bipyridyl)Re(CO)$_3$Cl (1-1) catalyst as quenching material added on glass/ITO substrate. (b) PL signal subtracted by the PL of the pure PVK reveals an isosbestic point at 485 nm (c) Stern-Volmer Plot of the data represented in (a) where PL^0 is the photoluminescence intensity of the pure PVK polymer without quencher and PL is the photoluminescence when a certain concentration of the quencher is added.

Figure 9.14(c) shows the Stern-Volmer Plot of the data represented in the Figure 9.14(a) before where PL^0 is the photoluminescence intensity of the pure PVK polymer without quencher at its maximum at approx. 410 nm and PL is the photoluminescence intensity when a certain concentration of the quencher is added. The last curve from Figure 9.14(a) with an amount of quencher material of $15 \cdot 10^{-6}$ g is not represented since the PL signal is almost negligible and hence difficult to evaluate. Furthermore, the inhomogeneity of the film increases with increasing rhenium catalyst concentration leading to lower PL signal of the catalyst acceptor at high concentrations.

According to the Stern-Volmer equation 9.4, a plot of (PL⁰/PL)-1 vs. quencher concentration should give a straight line which is observed for this system to a high degree (R-square value of 0.9967).[141]

$$\frac{P^0}{P} = 1 + K_{SV}[Q] \qquad (9.4)$$

In equation 9.4, K_{SV} is the Stern-Volmer constant and Q is the quencher concentration. Evaluating equation 9.4 for the data represented in Figure 9.14(c) results in a quenching constant K_{SV} of $2.06(\pm 0.08) \cdot 10^6 \, g^{-1}$. In the case of dynamic quenching, in e.g. Förster resonance energy or Dexter type electron transfer, the constant K_{SV} is the product of the true quenching constant k_q and the excited state lifetime τ in the absence of the quencher material, according to equation 9.5.

$$K_{SV}[Q] = k_q \cdot \tau_F \qquad (9.5)$$

The fluorescence lifetime of PVK (τ_F) has been reported with $5.34(\pm 0.03)$ ns. [142] Evaluating equation 9.5 for a K_{SV} of $2.06(\pm 0.08) \cdot 10^6 \, g^{-1}$ and a τ_F of $5.34(\pm 0.03)$ ns results in a quenching constant k_q of $3.85(\pm 0.17) \cdot 10^{14} \, g^{-1} s^{-1}$. This quenching constant k_q is then interpreted as the bimolecular reaction rate constant for the elementary reaction of the excited state with the quencher material Q.

Figure 9.15 shows PL quenching experiments with PVK as donor polymer and two different concentrations of (2,2' bipy)Re(CO)$_3$Cl (1-1) catalyst as quenching material added. The samples were dropcast from dichloromethane (DCM) solution. The PL spectra are corrected by the absorption of each film taking an average value of 10% for the contribution due to light scattering into account. The PVK concentration in all samples is $5 \cdot 10^{-4}$ g. The solid films are excited at two different wavelength namely 350 nm (black lines), which is the maximum of the PVK donor and relative minimum of the catalyst acceptor, and at 400 nm (red lines), which is the maximum of the ³MLCT catalyst acceptor absorption, respectively. The experiment is depicted for two different concentrations of the catalyst acceptor, $1 \cdot 10^{-6}$ g (solid lines) and $6 \cdot 10^{-6}$ g (dashed lines). In the experiment with the lower concentration, by exciting

9.3. POLYVINYLCARBAZOLE WITH (2,2'-BIPY.)RE(CO)$_3$CL

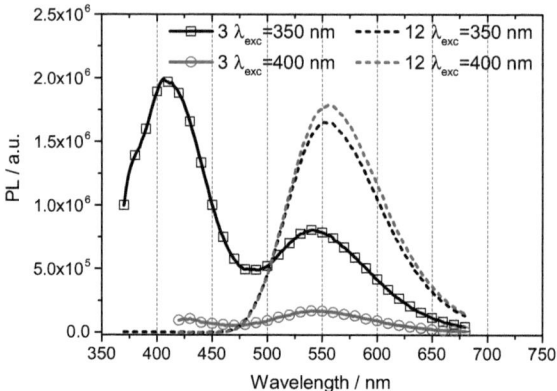

Figure 9.15: PL spectra with PVK as donor polymer and $1 \cdot 10^{-6}$ g (solid lines) or $6 \cdot 10^{-6}$ g (dashed lines) of (2,2'-bipyridyl)Re(CO)$_3$Cl (1-1) catalyst as quenching material on glass/ITO substrate. The material was dropcast from solution, the spectra are corrected to their corresponding absorption. The PVK concentration in all samples is $5 \cdot 10^{-4}$ g. The solid films are excited at two different wavelength namely 350 (black lines) and 400 nm (red lines) respectively.

at the polymer absorption maximum at 350 nm, two luminescence peaks are observed, one characteristic for the PVK donor polymer at about 400 nm and the other at about 550 nm, characteristic for the luminescence of the catalyst acceptor 1-1. When the sample is excited at 400 nm, namely at the ^{3}MLCT maximum o the acceptor 1-1, no PL signal of the donor polymer is present and only a weak PL signal of the catalyst material can be observed. At high catalyst quencher concentrations (dashed lines), independent of the excitation wavelength, no PL signal of the PVK donor polymer is present anymore, however for both wavelengths a strong PL signal of the acceptor catalyst 1-1 is observed. Following the argumentation of this results, together with the fact that for this system no light induced ESR signal could be observed, the process seems to follow a Förster- and/or Dexter type of energy transfer from PVK to (2,2' bipy)Re(CO)$_3$Cl (1-1) with only a minor contribution of charge transfer involved.

CHAPTER 9. ORGANIC SEMICONDUCTORS FOR CO_2 REDUCTION

9.3.2 Photoluminescence quenching in ACN solution

For the potential application of CO_2 reduction to a synthetic fuel by a photoinduced energy transfer from PVK as donor material to (2,2' bipy)Re(CO)$_3$Cl (1-1) as catalyst acceptor, bulk solid donor-acceptor mixtures are not applicable. In typical applications reported in literature, the CO_2 substrate is provided by purging an organic solvent (DCM, ACN, etc.) until gas saturation and dissolving the catalyst material to form a homogeneous solution mixture.[49, 113, 143]

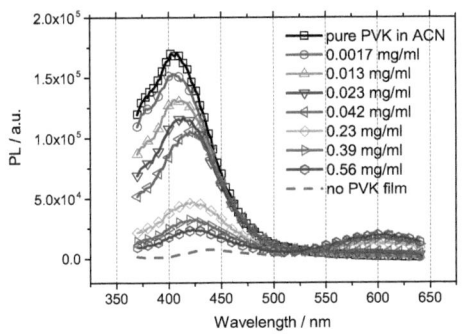

(a) PL quenching experiments with PVK

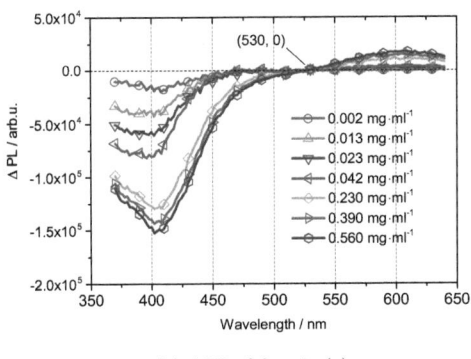

(b) ΔPL of data in (a)

Figure 9.16(a) shows PL quenching experiments with a solid, approx. 9nm thick film of PVK on ITO/glass substrate as donor polymer immersed into an acetonitrile (ACN) solution. The black solid line with squares shows the PL

9.3. POLYVINYLCARBAZOLE WITH (2,2'-BIPY.)RE(CO)$_3$CL

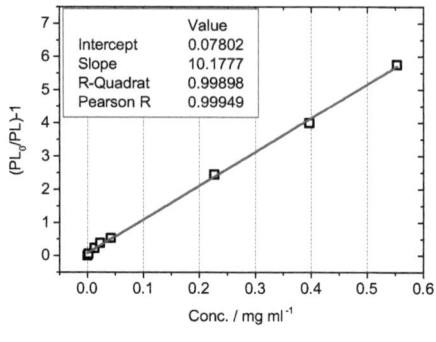

(c) Stern-Volmer Plot

Figure 9.16: (a) PL quenching experiments with a solid, approx. 9nm thick film of PVK on ITO/glass substrate as donor polymer and different concentrations of (2,2'-bipyridyl)Re(CO)$_3$Cl (1-1) catalyst as quenching acceptor added into the ACN solution. (b) PL signal subtracted by the PL of the pure PVK reveals an isosbestic point at 530 nm (c) Stern-Volmer Plot of the data represented in (a) where PL^0 is the photoluminescence intensity of the pure PVK polymer film without quencher present, and PL is the photoluminescence when a certain concentration of the quencher is added into the ACN solution.

signal of the pure PVK substrate in ACN excited at 350 nm. In the subsequent spectra (2,2' bipy)Re(CO)$_3$Cl catalyst was added as quenching material into the ACN solution. Following the increasing concentration of the catalyst material, clear quenching is observed with a similar behavior as previously shown for the bulk, solid donor-acceptor mixtures. The catalyst photoluminescence with its maximum at 553 nm appears, while the PVK luminescence with its maximum at about 410 nm vanishes almost completely once a catalyst concentration of 0.56 mgml^{-1} is reached (violet curve with circles). The gray dashed line shows the PL signal at the end of the quenching experiment when the PVK film was removed from the ACN solution. By looking at the spectra recorded in Figure 9.16(a) one has to take into account that only a few surface mono layers of donor and acceptor molecules get the chance to interact, while the bulk of the material in the solid phase and dissolved in solution will not take part in the actual energy transfer reaction. Following this argumentation it is remarkable observing a strong PL quenching effect in this solid-liquid donor-acceptor system. In Figure 9.16(b) the PL signal of the data in (a) is subtracted by the PL of the pure PVK. This plot reveals a isosbestic point at 530 nm as indicated in the graph, which is less pronounced and shifted to longer wavelength compared to the solid-solid mixture shown before (Figure 9.14(b), isosbestic point at 485 nm).

In Figure 9.16(c) the data represented in the Figure 9.16(a) is depicted in a Stern-Volmer Plot, where PL^0 is the photoluminescence intensity of the pure PVK polymer film without quencher present, and PL is the photoluminescence when a certain concentration of the quencher is added into the ACN solution. The plot of $(PL^0/PL)-1$ vs. quencher concentration gives again an almost perfect straight line (R-square value of 0.999). Evaluating equation 9.4 for the data represented in Figure 9.16(b) results in a quenching constant K_{SV} of $10.17(\pm 0.13) \cdot 10^3$ ml g^{-1}. Evaluating again equation 9.5 for K_{SV} of $10.17(\pm 0.13) \cdot 10^3$ ml g^{-1} and τ_F of $5.34(\pm 0.03)$ ns results in a quenching- or bimolecular reaction rate constant k_q of $1.90(\pm 0.04) \cdot 10^{12}$ ml g^{-1}s^{-1}.

Summarizing this work, the photoinduced energy transfer from poly(N-vinylcarbazole), as donor material, to fac-(2,2'-bipyridyl)Re(CO)$_3$Cl, as catalyst acceptor, could be shown for the first time. Photoluminescence quenching experiments revealed dynamic quenching due to resonance energy transfer in solid donor/acceptor mixtures and in solid/liquid systems. The bimolecular reaction rate constants for the elementary reaction of the excited state with the quencher material could be determined with $3.85(\pm 0.17) \times 10^{14}$ g^{-1}s^{-1} using Stern-Volmer analysis. This work shows that poly(N-vinylcarbazole) (PVK) as a decent an cheap absorber material can act efficiently as redox photosensitizer in combination with fac-(2,2'-bipyridyl)Re(CO)$_3$Cl as catalyst acceptor, which might lead to possible applications in photocatalytic CO_2 reduction. A detailed study of this work has been published in the Journal of *ChemPhysChem*, compare ref. nr. [144].

9.4 Functionalized Poly(3-hexylthiophene)

During the last 20 years electrically conducting and semiconducting polymers have received a great deal of attention due to their interesting properties and potential applications.20 Among a number of conducting polymers, polythiophene (PT) has attracted excessive interest since it is relative stable to both oxygen and moisture and can readily be used in electrochemistry.[145, 146, 147]

Therefore, the aim of this work is to functionalize poly (3-hexylthiophenes)

9.4. FUNCTIONALIZED POLY(3-HEXYLTHIOPHENE)

(P3HT) with fac-(2,2'-bipyridyl)Re(CO)$_3$Cl (1-1) for its potential application of electro- and photochemical CO_2 reduction. The monomer is deposited by electro-polymerization onto the electrode and used to determine its potential for heterogeneous catalysis towards CO_2 reduction. This interesting approach lead to the formation of a conducting polymer with functionalised rhenium bipyridine catalysts attached to the polymer backbone via alkyl bridges.

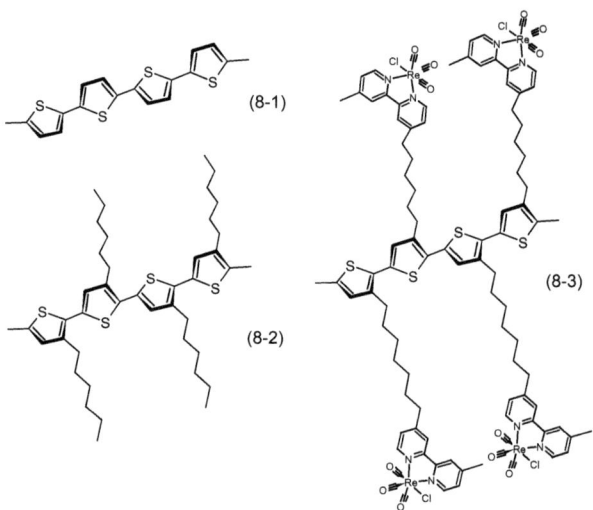

Figure 9.17: Schematic representation of three different conducting polymers investigated in this study, polythiophene (8-1), poly (3-hexylthiophene) (8-2) and fac-(2,2'-bipyridyl)Re(CO)$_3$Cl functionalized poly 3-(7-(4, 4'-methyl-(2,2'-bipyridil)ReCO$_3$Cl)heptyl)thiophene (8-3). (Materials were prepared by G. Aufischer.)

Figure 9.17 shows the schematic representation of three different conducting polymers investigated in this study, namely polythiophene (8-1), poly (3-hexylthiophene) (8-2) and fac-(2,2'-bipyridyl)Re(CO)$_3$Cl functionalized poly 3-(7-(4, 4'-methyl-(2,2'-bipyridil)ReCO$_3$Cl) heptyl) thiophene (8-3). This three different type of polymers are what is described in literature as three different generations of conducting polymers. Polythiophene (8-1) is a conjugated polymer with good conductivity, but with the disadvantage of low processability and solubility. To overcome this disadvantages P3HT (2) is a polyconjugated material with additional hexyl substitutions leading to a high solubility and processability without losing the interesting physical and chemical properties of the material. Finally the novel fac-(2,2'-bipyridyl)Re(CO)$_3$Cl functionalized P3HT (8-3) is a fully functionalized polyconjugated material with improved and new physical and chemical properties for the specific applications of electro- and photochemical CO_2 reduction. (The monomer synthesis is described in detail in the Bachelor thesis of G. Aufischer.)

9.4.1 Electropolymerisation and characterization

Electroactive monomers such as the monomer thiophene unit can polymerize to form conducting or semi conducting polymers. Such conducting polymers can be synthesized either chemically by using an external reactant or electrochemically by electropolymerization. The later has several advantages for deposition on conducting supporting electrodes compared to chemical polymerization and was the matter of choice in this work.

Electropolymerization is a well-known technique and the method mainly used for deposition of organic conducting or semiconducting films on conducting substrate electrodes like platinum. The procedure has several advantages, the most important ones are the possibility to control the rate of polymer nucleation and hence the growth of the film by proper selection of the electropolymerization parameters. This allows also the control of film thickness by the amount of charge that passed during the deposition process. Furthermore a suitable selection of solvent and supporting electrolyte allow to a certain degree control over the film morphology.[148] In this work, electropolymerization is performed under potentiodynamic conditions which was shown to result in the formation of electrochemically conducting polymer matrixes forming disordered spatial chains.[149]

Figure 9.18 shows the potentiodynamic electropolymerization of 0.1 M thiophene to polythiophene (8-1) in ACN solution containing 0.1 M TBAPF$_6$ as supporting electrolyte. The successive film formation is indicated by an increase in the current density of each cycle. The first scan is depicted in red line with circles and the last scan in blue line with triangles respectively. After electropolymerization the Pt working electrode was fully covered with a continuous shiny red to brown film as can be seen in the picture in Figure 9.18(b).

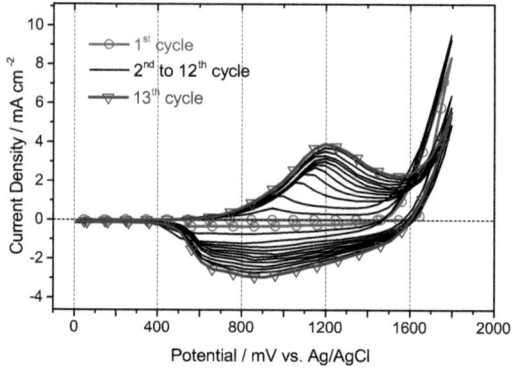

(a) Poythiophene electropolymerization (b) Picture of the electropolymerized film

Figure 9.18: (a) Potentiodynamic electropolymerization of 0.1 M thiophene to polythiophene (8-1) in ACN solution containing 0.1 M TBAPF$_6$ as supporting electrolyte. First scan (red line with circles) and last scan (blue line with triangles). The cyclic voltammograms were recorded at a scan rate of 100 mVs^{-1} inside the glove box using a Pt working electrode (WE), a Pt counter electrode (CE) and a Ag/AgCl quasi reference electrode. (b) Picture of the electropolymerized polythiophene on a Pt foil supporting working electrode.[Prepared by G. Aufischer]

CHAPTER 9. ORGANIC SEMICONDUCTORS FOR CO_2 REDUCTION

The mechanism for the polymerization has been reported to proceed via a radical coupling mechanism resulting in α-α linkages. Following the progress of the polymerization in Figure 9.18 one can observe a shift to lower oxidation potentials upon increasing scan numbers. This trend can be understood by the fact that the α-α linkages are a combination of monomer and oligomer radicals where the polymerization reaction occurs at lower potential on existing polymer than on a bare metal surface.[150] The cyclic voltammograms were recorded at a scan rate of $100\,mVs^{-1}$ inside the glove box using a Pt working electrode (WE).

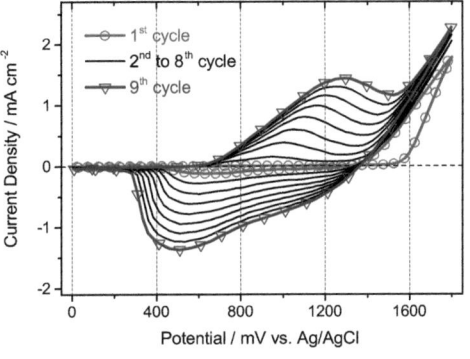

(a) Poly(3-hexylthiophene) electropolymerization

(b) Picture of the electropolymerized film

Figure 9.19: (a) Potentiodynamic electropolymerization of 0.056 M 3-hexylthiophene to poly (3-hexylthiophene) (8-2) in propylene carbonate solution containing 0.1 M $TBAPF_6$ as supporting electrolyte. First scan (red line with circles) and last scan (blue line with triangles). The cyclic voltammograms were recorded at a scan rate of $50\,mVs^{-1}$ inside the glove box using a Pt working electrode (WE), a Pt counter electrode (CE) and a Ag/AgCl quasi reference electrode. (b) Picture of the electropolymerized poly (3-hexylthiophene) on a Pt foil supporting working electrode.[Prepared by G. Aufischer]

Figure 9.19 shows the potentiodynamic electropolymerization of 0.056 M 3-hexylthiophene to poly (3-hexylthiophene) (2) in propylene carbonate solution containing 0.1 M $TBAPF_6$ as supporting electrolyte. The successive film formation is indicated by an increase in the current density of each cycle. The first scan is depicted in red line with circles and the last scan in blue line with triangles respectively. After electropolymerization the Pt working electrode was fully covered with a continuous dull brownish film as can be seen in the picture in Figure 9.19(b).

The mechanism for the polymerization has been reported to proceed via a radical coupling similar to the case of the polythiophene described before. Also, following the progress of the polymerization in Figure 9.19 one can observe again a shift to lower oxidation potentials upon increasing scan numbers. The cyclic voltammograms were recorded at a scan rate of $50\,\text{mVs}^{-1}$ inside the glove box using a Pt working electrode (WE), a Pt counter electrode (CE) and a Ag/AgCl quasi reference electrode.

Comparing the electrochemical polymerization of poly (3-hexylthiophene) (8-2) in Figure 9.19 to the polythiophene (8-1) in Figure 9.18, it can be stated that the conductivity of the film does not appear to be influenced significantly by the presence of the alkyl substituents on the 3-position although one might expect a structural change in the film due to steric interactions which could result in lower conductivities. Generally it has been reported, that for polythiophene (8-1) as well as for poly (3-hexylthiophene) (8-2) conductivities are in the order of 1-100 S cm^{-1}.[148, 151, 152]

Electropolymerization of Re(4-methyl-4'-(7-thienylheptyl)-2,2'-bypyridene-(CO)$_3$Cl to fac-(2,2'-bipyridyl)Re(CO)$_3$Cl functionalized poly (3-hexylthiophene) (8-3) as the representative compound for the so called 3rd generation type of conducting polymer was difficult in the sence that the substituent fac-(2,2'-bipyridyl)Re(CO)$_3$Cl groups undergo irreversible oxidation above a potential of 1500 mVvs.NHE compare Figure 6.4 in chapter 6. One possible way to overcome this problem is to use BF$_3$–diethyl ether (BFEE) which substantially lowers the required potential for electro polymerization.[153]

Figure 9.20 shows the potentiodynamic electropolymerization of approx. 0.015 M Re(4-methyl-4'-(7-thienylheptyl)-2,2'-bypyridene(CO)$_3$Cl to fac-(2,2'-bipyridyl)Re(CO)$_3$Cl functionalized poly (3-hexylthiophene) (8-3) in in BFEE solution. The successive film formation is indicated by an increase in the current density of each cycle. The first scan is depicted in red line with circles and the last scan in blue line with triangles respectively. After electropolymerization as shown in Figure 9.20(a) the film on the Pt working electrode was barely visible. For illustrative purposes an additional, thick film has been grown in BFEE containing 5 vol.% PPC and otherwise identical conditions for the picture depicted in Figure 9.20(b). For the thick film formation in (b) it can be seen that

CHAPTER 9. ORGANIC SEMICONDUCTORS FOR CO_2 REDUCTION

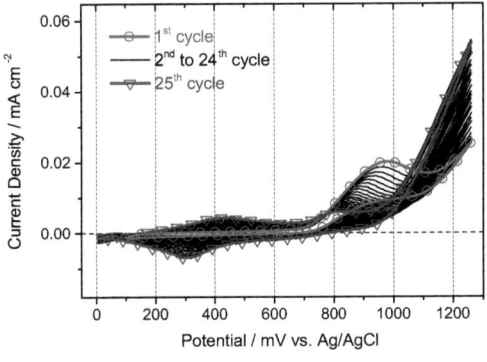

(a) Electropolymerization of (2,2'-bipyridyl)Re(CO)$_3$Cl functionalized poly (3-hexylthiophene)

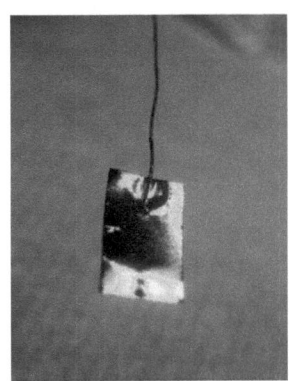

(b) Picture of the electropolymerized film

Figure 9.20: (a) Potentiodynamic electropolymerization of approx. 0.015 M Re(4-methyl-4'-(7-thienylheptyl)-2,2'-bypyridene(CO)$_3$Cl to fac-(2,2'-bipyridyl)Re(CO)$_3$Cl functionalized poly (3-hexylthiophene) (8-3) in in BFEE solution. First scan (red line with circles) and last scan (blue line with triangles). The cyclic voltammograms were recorded at a scan rate of $50\,\mathrm{mVs^{-1}}$ outside the glove box using a Pt working electrode (WE), a Pt counter electrode (CE) and a Ag/AgCl quasi reference electrode. (b) Picture of a different, very thick electropolymerized fac-(2,2'-bipyridyl)Re(CO)$_3$Cl functionalized poly (3-hexylthiophene) (8-3) on a Pt foil supSporting working electrode.[Prepared by G. Aufischer]

the Pt working electrode was fully covered with a continuous, almost gold like, shiny film. For further electrochemical characterization and electrochemical CO_2 reduction experiments the film formed in the experiment shown in Figure 9.20(a) was used due to its superior performance.

9.4.2 Cyclic voltammetry studies on electrochemical CO_2 reduction

Once the electroactive polymer electrodes were grown the prepared electrodes could be transfered into a monomer free electrolyte solution for their characterization towards electrochemical CO_2 reduction. Figure 9.21(a) shows the electrochemical characterization of the electropolymerized poly (3-hexylthiophene) (8-2) electrode deposited in the experiment depicted in Figure 9.20 before. The characterization is carried out for negative bias in in nitrogen- (black line with squares) and CO_2- saturated electrolyte solution (red line with circles), respectively. For the poly (3-hexylthiophene) (8-2) the reduction starts at about $-1700\,\mathrm{mVvs.Ag/AgCl}$ and is partly reversible with the re-oxidation maximum occurring at about $-1800\,\mathrm{mVvs.Ag/AgCl}$.

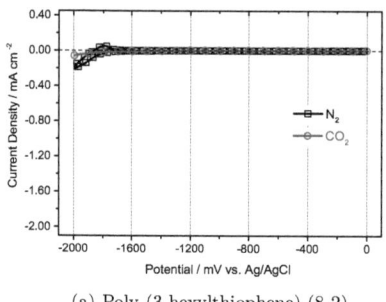
(a) Poly (3-hexylthiophene) (8-2)

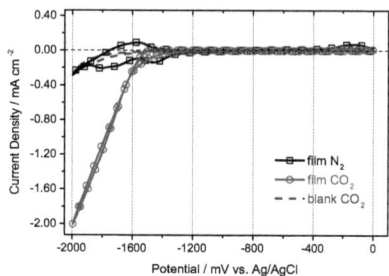
(b) fac-(2,2'-bipyridyl)Re(CO)$_3$Cl functionalized poly (3-hexylthiophene) (8-3)

Figure 9.21: (a) Cyclic voltammograms of poly (3-hexylthiophene) (8-2) electropolymerized on a Pt plate electrode and in (b) cyclic voltammograms of a thin film of fac-(2,2'-bipyridyl)Re(CO)$_3$Cl functionalized poly (3-hexylthiophene) (8-3) electrochemically polymerized on Pt plate electrode in nitrogen- (black line with squares) and CO_2- saturated electrolyte solution (red line with circles), respectively. The scan in the presence of CO_2 shows a large current enhancement for the polymer 8-3 due to the catalytic reduction of CO_2. A scan with no catalyst film present under CO_2 (blue dashed line) shows little to no reductive current. Voltammograms were recorded at $50\,\mathrm{mVs^{-1}}$ in a propylene carbonate solution with a concentration of 0.1 M TBAPF$_6$ as supporting electrolyte. A Pt working electrode (WE) with and without film, a Pt counter electrode (CE) and a Ag/AgCl quasi reference electrode.[Prepared by G. Aufischer]

It is important to notice, that in presence of CO_2 the magnitude of the reductive current is much lower compared to that of the curve obtained under inert N_2 saturated electrolyte solution with no characteristic features of the poly (3-hexylthiophene) (8-2) present any more. A similar behavior has been observed at our institute for a different class of organic semiconducting materials, namely Quinacridone (QNC). It was found that upon purging with CO_2 the characteristic reduction peak of QNC at $-1480\,\mathrm{mVvs.Ag/AgCl}$ diminished and the cathodic current decreased substantially showing similar features than represented in Figure 9.21(a). This phenomenon was related to the formation of a QNC-carbonate salt and could be used for efficient and controlled CO_2 capture and release. In the case of QNC it was shown that 20 w% capture (20g CO_2/100g QNC) could be achieved using controlled electrochemical reduce-and-capture. Since controlled capture, storage and release of CO_2 is a key step for both sequestration and utilization approaches of CO_2 this findings about

CHAPTER 9. ORGANIC SEMICONDUCTORS FOR CO_2 REDUCTION

organic semiconducting materials may become particularly interesting in the near future.

Figure 9.21(b) shows cyclic voltammetry measurements of a thin film of fac-(2,2'-bipyridyl) $Re(CO)_3Cl$ functionalized poly (3-hexylthiophene) (8-3) electrochemically polymerized on Pt plate electrode in nitrogen- (black line with squares) and CO_2- saturated electrolyte solution (red line with circles), respectively. This measurement constitutes the same thin film as represented in the measurement depicted in Figure 9.20(a). The solutions were purged with N_2 (black curve) or CO_2 (red curves) under stirring for about 15 minutes before cyclic voltammograms were recorded inside the glove box having O_2 and H_2O levels in the ambient atmosphere below 10 ppm. The measurement of the potential window with two Pt electrodes as WE and CE, respectively, and an electrolyte solution under N_2 does not show any reductive current in the potential range from 0 to -2000 mVvs.Ag/AgCl. When the solution is purged with CO_2 and no polymerized catalyst film is present, a minor reductive current starts to flow at a potential lower than about -1800 mVvs.Ag/AgCl (blue dashed line). When the Pt working electrode is replaced by the thin film of fac-(2,2'-bipyridyl)$Re(CO)_3Cl$ functionalized poly (3-hexylthiophene) (8-3) however, and measured in the electrolyte solution under N_2 atmosphere, the typical reduction curve of the polymer (8-3) is measured showing low current densities upon a bias to -2000 mVvs.Ag/AgCl with current densities of about -0.25 mAcm^{-2}. If however the electrolyte solution is at CO_2 saturation, a high, non-reversible reductive current enhancement is observed (Figure 9.21(b), red line with circles). The reductive current begins to increase at about -1550 mVvs.Ag/AgCl reaching current densities of about -2 mAcm^{-2} at potentials of -2000 mVvs.Ag/AgCl. Although a direct proof of reduction products is still missing, the current enhancement can be understood by a catalytic reduction of CO_2 to CO by the functionalized (2,2'-bipyridyl)$Re(CO)_3Cl$ catalyst on the alkyl substituents of the polythiophene. It is also interesting to mention that the voltammogram under CO_2 saturation shows peak crossing which is an inherent characteristic of the homogenous (2,2'-bipyridyl)$Re(CO)_3Cl$ catalyst as discussed before, compare Figure 6.6 and 6.7 in chapter 6 of this thesis.

Concluding this work it was shown for the first time the possible application of a functionalized poly (3-hexylthiophenes) with fac-(2,2'-bipyridyl)$Re(CO)_3Cl$, as a so called third generation of conducting polymer, for its application of electrochemical CO_2 reduction. Upon CO_2 saturation the cyclic voltammograms showed a strong current enhancement which is a good evidence for a catalytic reduction of CO_2 to CO by the functionalized (2,2'-bipyridyl)Re(CO)$_3$Cl catalyst. Beside its apparent application for CO_2 reduction, simple poly (3-hexylthiophene) might have interesting properties for controlled CO_2 capture and release and will be investigated further.

Chapter 10

Summary and future studies

In this work the photoinduced electron transfer from organic semiconductors onto redox mediator catalysts for CO_2 reduction has been investigated, starting with the identification, characterization and test of suitable catalyst materials. Once the catalyst redox mediators were found the next step was the immobilization of the catalyst on the electrode and thereby changing from homogeneous to heterogeneous catalysis. Finally, the combination of catalysts with organic semiconductor materials was investigated for energy and charge transfer from the donor polymer to the catalyst acceptor towards its possible applications in photocatlytic CO_2 reduction.

10.1 What has been accomplished

After the introduction chapter, where the scope and motivation of this work were defined, the thesis starts with an exhaustive experimental description in chapter 2,
"Experimental techniques", describing in detail the methods used for the experimental results obtained herein. The following chapter 4, *"Photophysical results"*, studies in great depth the photophysical properties of the catalyst materials further investigated for CO_2 reduction within the thesis.

In the subsequent chapter 5 of this work, *"Quantum chemical calculations"*, quantum chemical calculations were successfully used for the interpretation of experimental obtained data, such as infrared absorption spectra, and for the calculation of molecular orbital energy levels. Infrared spectroscopy allowed analytical characterization of molecular substances. However, since the vibrational degrees of freedom for non linear molecules scales with $3N$-6, with N being the number of atoms in the molecule, infrared spectra are complicated for larger molecules making the spectra difficult to interpret. For this reason calculated IR absorption spectra by quantum mechanical DFT calculations were successfully used to correlate characteristic features in the measured spectra to their molecular origin. Experimentally observed data and quantum chemical predictions for the IR spectra of the novel compound are in good agreement. For the prediction of measured infrared spectra it was sometimes necessary to scale the obtained calculated data by a constant factor to achieve good matching of calculated and measured spectra. Additionally quantum mechanical calculations were carried out for the determination of molecular orbital frontier energy levels. Calculations gave typically a lager band gap than band gaps estimated by UV-Vis absorption or cyclic voltammetry measurements with offsets of about 0.2 eV. It should be notice that quantum mechanical calculations are based on many approximations giving only an estimate of the real molecular properties.

In the following chapter 6 the materials investigated for homogeneous electro catalysis were Rhenium diimine complexes with different ligand systems as well as the Pyridinium catalyst. In the three sections *"Rhenium compounds with bipyridine ligands"*, *"Rhenium compounds with bian ligands"* and *"Pyridinium as catalyst"* the properties of these materials as homogeneous electro catalysts for CO_2 reduction were investigated. Figure 10.1 shows a summary of the materials investigated within this work and the obtained faraday and energy efficiencies for electrocatalyic CO_2 reduction and the quantum yield for photocatalytic CO_2 reduction.

Several tricarbonylchlororhenium(I)pyridyl complexes (1-1 to 1-4) were studied regarding to their potential as catalysts for homogeneous electrochemical and photochemical reduction of CO_2 to CO. It was found that the CO_2 reduction potential, determined by cyclic voltammetry, can be positively influenced by a modified ligand system. A comparison between the catalyst material 1-3 with added phenylethynyl groups at the 5,5' position to the non-modified catalyst 1-1 showed a shift in the onset CO_2 reduction potential by about 300 mV (vs. NHE) to more positive values. Bulk CO_2 electrolysis experiments showed Faraday efficiencies around 45 to 50 % for the formation of CO. The

CHAPTER 10. SUMMARY AND FUTURE STUDIES

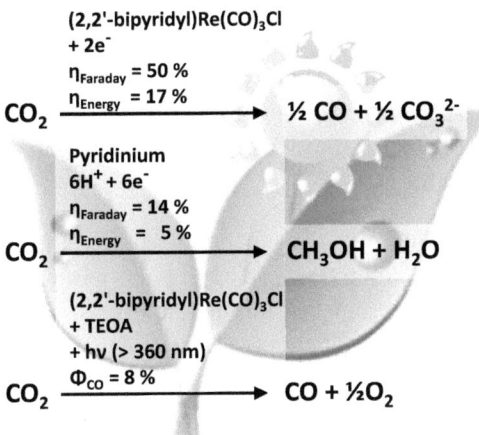

Figure 10.1: Summary of the materials investigated within this work and the obtained efficiencies. Faraday efficiency $\eta_{Faraday}$, energy efficiency η_{Energy} and quantum yield Φ_{CO} for CO_2 reduction.

best estimated rate constants according to cyclic voltammetry data are in the order of $220\,\text{M}^{-1}\text{s}^{-1}$ for compound 1-3.

The electrocatalytic properties of rhenium(I) tricarbonyl complexes carrying bis(arylimino)acenaphthene (BIAN) ligands for the selective two-electron reduction of CO_2 to CO in homogeneous solution have been successfully tested and characterized. It could be demonstrated that a variation of the ligand substitution pattern in close proximity to the metal center has a very significant influence on the catalytic performance of these systems. Further studies on the suitability of this deeply colored and readily tunable class of compounds as functional components of photocatalytic CO_2-reduction cycles are currently underway.

As a next step new and improved electro- and photocatalysts for CO_2 reduction have to be developed and characterized. The modification of the attached ligand system is a suitable way for systematic tuning of the excited state properties of such materials and hence their electro- and photocatalytic abilities. A clear demonstration of this was shown by the comparison of compound 1-1 and 1-3. Following this idea a modification where the phenylethynyl groups are attached to the 4,4' position might be highly interesting, assuming a better conjugation to the metal center of the complex. Although further catalyst improvements are crucial, finally, the two systems capable of catalytic CO_2 reduction and H_2O splitting could be combined in a photo- or photoelectrochemical cell. Such a system would be capable of solar powered production of syngas and its equivalents (CO and H_2 or NADH).

One more example of homogeneous catalysis investigated in chapter 6 was the pyridinium-catalyzed reduction of CO_2 to methanol. Cyclic voltammetry experiments were used to proof the important role of platinum in the overall reaction mechanism. Additionally controlled potential electrolysis were carried out over a extended period of 30 hrs leading to the formation of methanol via the pyridinium-catalyzed reduction of CO_2.

Additionally, some of these materials proofed to be efficient photo catalysts for the reduction of CO_2 producing mainly CO with the aid of a sacrificial electron donor as for example triethanolamine (TEOA). For photo-catalysis, it is important to extend the absorption of the catalyst compounds in the visible region. In this work new catalysts 1-3 and 1-5 were synthesized which show significantly higher absorption in the visible range, compare chapter 7. It was assumed to be a clear benefit for photocatalytic application, however, it was shown that the experiments over several hours of irradiation yielded only very low CO formation. Under identical conditions, the unmodified compound 1-1 still performs better for photocatalytic CO_2 reduction. To the best of our understanding, following quantum mechanical DFT calculations, this different behaviour might be attributed to an inversion of the lowest-lying excited state properties of the new compound compared to the situation in compound 1-1, which is crucial for the photochemical reactivity of such systems.

In chapter 8, *"Heterogeneous electro catalysis"*, several ways to change from homogeneous to heterogeneous catalysis for CO_2-reduction were studied. As one of the first trials, the rhenium catalyst 1-2, with dicaboxylic side groups at the ligand system, was used to attach to a binding metal oxide layer. Although in this experiments the anchoring to the oxide layer was successfully achieved, it tourned out that the film was not stable upon electrochemical reduction. The cyclicvoltammetry experients showed, that below a potential of about $-1000\,mV$ vs. NHE the layer was irreversibly destroyed and rinsed of the electrode as a black precipitate.

CHAPTER 10. SUMMARY AND FUTURE STUDIES

The electropolymerization of the active catalyst compound 1-3 onto a Pt working electrode is investigated. The electropolymerization to the electrode surface and its potential for heterogeneous catalysis towards CO_2 reduction was successfully studied. The film growth on a Pt working electrode was performed by potentiodynamic reductive scanning in nitrogen saturated acetonitrile solution. The films were electrochemically characterized using cyclic voltammetry. The catalytic properties for CO_2 reduction were studied via cyclic voltammetry in carbon dioxide saturated acetonitrile solution. It was found that the electropolymerisation of the monomer 1-3 appears to be a promising way to change from homogeneous to heterogeneous catalysis for CO_2-reduction, reducing the amount of required catalyst material needed and overcoming the limitations regarding solubility of homogeneous catalysts in general. The novel catalyst film furthermore demonstrates high selectivity for the CO_2-reduction to CO at relatively low reduction potentials and high current densities. Figure 10.2 shows the schematic representation of the differences between a homogeneous and a heterogeneous process for catalytic CO_2 reduction

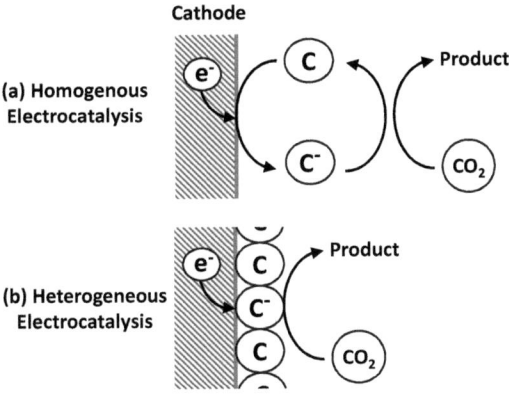

Figure 10.2: Schematic representation of the differences between a homogeneous and a heterogeneous process for catalytic CO_2 reduction

Finally, following a different approach, the catalyst material (2,2'-bipyridyl)-Re(CO)$_3$Cl (1-1) was immobilized into a polypyrrole matrix by electrochemical polymerization of pyrrole in a homogeneous mixture of pyrrole and the catalyst material. Pyrrole was electropolymerized on a Pt foil serving as supporting working electrode for the polypyrrole film. First experiments using cyclic voltammetry looked promising, however no product gas analysis was performed

10.1. WHAT HAS BEEN ACCOMPLISHED

to independently verify the CO_2 reduction and confirm the expected CO formation. Subsequent experiments failed in reproducing this data and further studies on this effect are under current investigations.

In the last chapter 9, *"Organic semiconductors for CO_2 reduction"*, the combination of catalysts with organic semiconductor materials were investigated. In the first part poly(3-hexylthiophene) (P3HT) was used as the organic semiconducting polymer which served as donor material and Pyridinium as a homogeneous electrocatalyst in solution as acceptor. In the photoresponse of a P3HT covered ITO electrode at constant potential upon white light irradiation and a varying pyridinium concentration it was found, that when the catalyst concentration is increased, the corresponding current photo-response is enhanced. This is expected when pyridinium acts as an efficient acceptor molecule. Bulk electrolysis experiments were performed, however no products in form of methanol could be detected. This can be explained since the lifetime of this type of electrodes is rather low and the photocurrent is in the μA regime. In the case with the P3HT electrodes the electrodes corroded within 2 to 3 hrs of electrolysis time. As a resulting conclusion, this system has to be improved significantly in terms of electrode stability and current densities which is subject to ongoing research.

In a different approach, poly(N-vinylcarbazole) (PVK) was used as absorber material acting as efficient redox photosensitizer in combination with fac-(2,2'-bipyridyl)Re(CO)$_3$Cl (1-1) as catalyst acceptor. The energy levels of the donor polymer and acceptor catalyst are aligned in a favorable situation for photo-excited charge and/or energy transfer. Subsequent photoluminescence (PL) quenching and light induced ESR experiments were carried out to study this system in bulk, solid phase mixtures of donor and acceptor, and at a donor-acceptor solid-liquid interface. PL quenching experiments with PVK as donor polymer in solid film mixtures and in a solid-liquid interface between polymer and catalyst revealed strong quenching due to a Förster- and/or Dexter type of energy transfer from PVK to (2,2' bipy)Re(CO)$_3$Cl (1-1) with only a minor contribution of charge transfer involved. This findings might leading to possible applications in photocatlytic CO_2 reduction.

The last part of this thesis investigated a method to functionalize poly (3-hexylthiophenes) (P3HT) with fac-(2,2'-bipyridyl)Re(CO)$_3$Cl (1-1) for its potential application of electro- and photochemical CO_2 reduction. The monomer was deposited by electro-polymerization onto the electrode and used to determine its potential for heterogeneous catalysis towards CO_2 reduction. This interesting approach lead to the formation of a conducting polymer with functionalised rhenium bipyridine catalysts attached to the polymer backbone via alkyl bridges. It was shown for the first time the possible application of a functionalized poly (3-hexylthiophenes) with fac-(2,2'-bipyridyl)Re(CO)$_3$Cl, as a so called third generation of conducting polymer, for its application of electrochemical CO_2 reduction. Upon CO_2 saturation the cyclic voltammograms showed a strong current enhancement which is a good evidence for a catalytic reduction of CO_2 to CO by the functionalized (2,2'-bipyridyl)Re(CO)$_3$Cl catalyst. Beside its apparent application for CO_2 reduction, simple poly (3-hexylthiophene) might have interesting properties for controlled CO_2 capture and release, which will be investigated further.

10.2 The need of novel catalysts

The fac-(2,2'-bipyridyl)Re(CO)$_3$Cl (1-1) is up to now one of the most efficient catalysts for the reduction of CO_2 with reported quantum yields (Φ_{CO}) for photocatalytic CO_2 reduction of 0.14. Subsequent modifications of the bipyridine ligand system of 1-1 lead to the development of superior catalysts with Φ_{CO} up to 0.59, making these type of materials the most efficient CO_2 photo-catalyst among known homogeneous catalyst materials by now.[58, 111] One substantial drawback of this type of materials is based on the central metal atom rhenium being extremely rare and hence expensive. Rhenium is with an abundance of 0.7 ppb in the earth's crust[154] one of the rarest materials available resulting in a rhenium metal pellet price of 4630 $ per kilogram. [155]

The problems in using rhenium as catalytic metal center is obvious. The metal is rare and hence very expensive, therefore limiting its potential application for large scale CO_2 reduction for artificial fuel synthesis. Taking this into account, one of the major challenges for basic research is to find catalysts based on earth-abundant materials that have low over potentials for CO_2 reduction, water oxidation and a high turnover frequency (equivalten to the rate constants k). Manganese offers a promising alternative to rhenium in the M^{+1} (d^6) configuration. Manganese is about 1.3 million times more abundant in earth's crust than rhenium. However, manganese is a smaller atom than rhenium. This has consequences on the nature of the LF and MLCT states. When manganese is used instead of rhenium, the very low energy CO stretching frequencies indicate that the metal-CO back bonding (π-acceptor) is significantly increased as compared to the rhenium complex.[156, 157] One very interesting approach has resently been published by Kubiak and others where rhenium was successfully substituted by manganese in the CO_2 reduction catalysts Mn(bpy-tBu)(CO)$_3$Br and Mn(bpy-tBu)(CO)$_3$(MeCN)(OTf).[156, 158] It is expected that this achievements will have substantial impact on the scientific field working on CO_2 reduction.

Another very interesting approach was presented by Bocarsly and his group as they've reported the reduction of carbon dioxide to methanol and formic acid on a platinum electrode in aqueous solutions containing pyridinium as active catalyst material.[31] Following ongoing scientific reports over the last two years, the proposed mechanism and the role of pyridinium as the catalyst material lead to a controversial discussion in the field,[107] the reported data is nonetheless impressive and deserves great attention by the scientific community.

It has been stated in the beginning of the introduction chapter 1 of the thesis that converting sunlight into synthetic chemical fuels by the direct reduction of CO_2 is probably the most elegant and sustainable way to overcome demanding problems related with the current energy supply of humanity. How important the role of high turnover frequencies in catalyst materials is, becomes apparent when we consider practical applications and look at photovoltaic current densities vs. turnover frequencies (TOF) of CO_2 reduction catalysts to date. The best photovoltaic materials can support current densities of 10 to 20 mA cm^{-2} (or 100 to 200 A m^{-2}). In contrast, for a two electron process, a monolayer of catalyst with a TOF of 1000 s^{-1}, will support a current density of only 0.1 mA cm^{-2}.

Figure 10.3 relates current densities to equivalent turnover frequencies of a monolayer of catalyst material. Given current technology in photovoltaic cells, typical photovoltaic current densities are in the order of 100 A m^{-2}. As indicated by the red circle in Figure 10.3, a monolayer of catalyst would need to achieve a TOF of about 100 000 s^{-1} to support the corresponding current density. The best catalyst materials known by now have about 500 - 1000 s^{-1}.

CHAPTER 10. SUMMARY AND FUTURE STUDIES

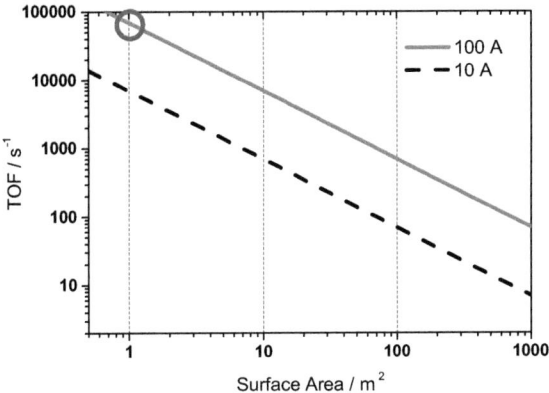

Figure 10.3: Relation between current densities vs. turnover frequencies of a monolayer of catalyst material for the two electron reduction of CO_2 to CO.

Obtaining high photocurrent densities inalienably requires the use of molecular catalyst with high TOFs. In theory one could push the system also by increasing the molecular catalyst concentration, however, higher catalyst concentrations in solution will lead to an increase in light absorption by the solution and the solubility of the catalyst is limited in any given solvent. Taking this into account, there is an apparent need to advance catalyst materials by identifying the rate limiting intermediate steps in CO_2 reduction and improve them.

10.3 A more elaborate view on CO_2 reduction to date

In this thesis the main focus was on the CO_2 reduction by the use of catalyst materials concentrating on the cathode reaction in the electrochemical setup only. Especially cyclic voltammetry experiments using a three electrode system were used in great depth to investigate the properties of various materials for CO_2 reduction. However, one must not forget that by using a three electrode setup only the potential drop on the working electrode is measured, which is one of the important parameters for determining the effectiveness of a catalyst material but not of the total cell. The question that remains is, how this potential is related to a fully working device capable of CO_2 reduction, a pro-

10.3. A MORE ELABORATE VIEW ON CO_2 REDUCTION TO DATE

totype electrolyser so to speak, being a device similar to the one depicted in the introduction chapter 1, compare scheme 1.4? In a real CO_2 electrolyser, if everything is hooked together, several potential losses have to be accounted for. Let's assume a CO_2 reduction following the reactions 10.1 on the cathode and 10.2 on the anode with CO and O_2 as main products.

$$CO_2 + 2H^+ + 2e^- \rightleftharpoons CO + H_2O \qquad (10.1)$$

$$H_2O \rightleftharpoons \frac{1}{2}O_2 + 2H^+ + 2e^- \qquad (10.2)$$

There is the well-studied potential for CO_2 reduction on the cathode electrode E_C, the potential E_A, that has to be applied on the anode counter electrode for the O_2 evolution reaction, the potential drop due to ohmic losses in the electrolyte solution E_O and the junction potential E_J over the membrane separating anode and cathode compartment. Taking all this potential drops into account, the total voltage which has to be applied, E_{total}, can be calculated by the following equation 10.3.

$$E_{total} = E_C + E_A + E_O + E_J \qquad (10.3)$$

The potential necessary for CO_2 reduction, E_C, is about -1.65 V vs. NHE taking the potential for the 2nd reduction wave of the catalyst material (2,2'-bipyridyl)Re(CO)$_3$Cl (1-1) as the standard catalyst and benchmark compound deeply investigated in this thesis, compare chapter 6.

For the oxygen evolution to occur, several catalyst materials have been reported in literature. One of the most promising and applicable is a cobalt–phosphate type of catalyst where the O_2 evolution is reported to occur at approximately +1.3 V vs NHE (E_A) in an aqueous phosphate buffer solution at pH 7. [159, 160]

The ohmic potential drop in the cell E_O depends mainly on the conductivity of the electrolyte solution used, the area of the electrodes and the distance between working and counter electrode. For a 0.1 M KCl solution in water the electrolyte conductivity is about 12.8 mS cm^{-1} at 25°C.[161] A typical distance between working and counter electrode is 2 cm with current densities of 2 mA cm^{-2}. Taking these values for an electrode of 10 cm^2 one obtains an ohmic potential drop in the cell E_O of about 312.5 mV.

The specific conductivity of a cation exchange membrane of 0.56 mm thickness used to separate anode and cathode compartment is about 9.3 mS cm^{-1} in a 0.1 M NaCl solution, resulting in a membrane potential drop E_J that is in the order of 50 mV for a current density of 2 mA cm^{-2} and hence almost negligible for low current densities.[162, 163, 164]

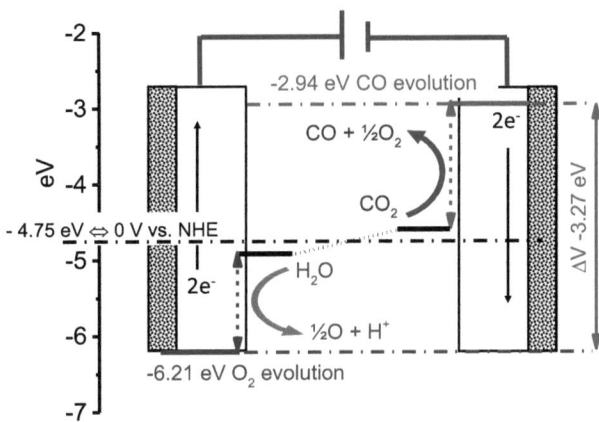

Figure 10.4: Schematic representation of the characteristic potentials necessary for a CO_2 electrolyser system consisting of two electrodes with a membrane to separated anode and cathode compartment.

Scheme 10.4 depicts the apparent voltage necessary for a CO_2 electrolyser system with a E_C of -1.65 V for the CO_2 reduciton, E_A of 1.3 V for the oxygen evolution, E_O of 312.5 mV for the ohmic losses in the electrolyte and E_J of 50 mV for the membrane potential. The total voltage that needs to be applied given by equation 10.3 for the actual system sums up to 3.27 V. One has however to take into account that the reduction potential E_{red} for the CO_2 reduction following reaction 10.1 with CO and O_2 as main reduction products is E_{red} -0.53 V and for the reaction 10.2 is E_{ox} 0.82 V (at pH 7 and 25 °C).

The sum of both potentials results in a thermodynamically necessary electro-

10.3. A MORE ELABORATE VIEW ON CO_2 REDUCTION TO DATE

motive force emf of 1.35 V for the formation of CO and O_2. Since the necessary applied voltage in our system is 3.27 V, the device efficiency can then be calculated by 1.35 V divided by 3.27 V giving an efficiency of about 41 %, assuming a Faraday efficiency of 100 %.

In the electrochemical CO_2 reduction process pure CO_2 is used. Capturing CO_2 from the air is particularly important since the atmospheric concentrations are relatively low with about 400 ppm or 0.04 % by the end of 2013.[3] If we assume a perfect mechanism for capturing of CO_2, only the free energy of mixing has to be provided, which can be calculated by knowing the partial pressure difference upon concentration according to equation 10.4.[165]

$$\Delta G° = R \cdot T \cdot \frac{P}{P°} \qquad (10.4)$$

Where P° is 1 bar atmospheric pressure, P is the partial pressure of CO_2 with $4 \cdot 10^{-4}$ bar, and T is the temperature taken at 300 K, which gives -19.51 kJ mol^{-1}. One must not forget that this is for a perfect absorber, in reality however, the energy needed is expected to be substantially higher.

The theoretical energy that can be gained back by burning CO and O_2 again according to reaction 10.5

$$CO + \frac{1}{2}O_2 \rightleftharpoons CO_2 \qquad (10.5)$$

has a change in the standard Gibbs energy of reaction ΔG of -257 kJ mol^{-1}. The Gibbs energy of reaction is also related to the electromotive force according to equation 10.6.

$$\Delta G = -n \cdot F \cdot E \qquad (10.6)$$

Taking now the total voltage applied in the CO_2 electrolysis system of 3.27 V, the change in Gibbs energy of reaction according to equation 10.6 is -631 kJ mol^{-1}, which has to be applied to the system for the CO_2 reduction reaction to occur, plus the free energy of mixing for concentrating the

CO_2 in the system with minimal about $-19.51\,kJ\,mol^{-1}$, so in total about $-651\,kJ\,mol^{-1}$. Based on these assumptions the overall CO_2 reduction to energy return efficiency would be $-257\,kJ\,mol^{-1}$ devided by $-651\,kJ\,mol^{-1}$, or about 38 %. Gasoline has a specific volumetric energy density of about 35 MJ/l (9.7 kWh/l).[7] Taking the calculated efficiency value of 38 % and a price for electric energy with 7.5 cent/kWh, the production of an energy equivalent of 1 l gasoline by CO_2 reduction to CO would cost about 1.9 Euro and would remove 3.4 m^3 CO_2 from the atmosphere, while the price for gasoline at a local gas station in Austria was about 1.4 Euro/l by the time this sentence was written.

Of course, these rough and quick calculations are by no means exhaustive and only schematic at best, but they might still give a glimpse that electro- and photochemical CO_2 reduction for artificial fuel synthesis might become at some point more to society than an interesting academic field of research only. I would like to finish this work with quoting the Nobel Laureate Wilhelm Ostwald (1853 - 1932), who was one of the first scientists that was deeply aware of the importance of energy storage and conversion and predicted many forthcoming developments [166]:

> *"When we ask for the general function of science, just as we do in the case of art, the answer is very straight forward: it consists of prophecies. All the manifold work driven by the sciences has the final goal to give us the potential to foresee future occurrences"*. [167]

10.3. A MORE ELABORATE VIEW ON CO_2 REDUCTION TO DATE

Chapter 11

Appendix

List of abbreviations

AC = alternating current

ACN = acetonitrile

AFM = atomic force microscopy

ATR = attenuated total reflectance

a.u. = arbitrary unit

BCB = benzocyclobutene

bipy = bipyridine

BLZ = blazing angle

CMS = charge modulation spectroscopy

DCM = dichloromethane

DFT = density functional theory

DMF = dimethylformamid

DTGS = triglycine sulfate

fac = facial

FTIR = fourier transform infrared spectroscopy

GC = gas chromatography

HF = Hartree-Fock

HOMO = highest occupied molecular orbital

IC = ion chromatography

IL = intra ligand

ITO = indium tin oxide

LUMO = lowest unoccupied molecular orbital

MCT = mercury cadmium telluride

MeOH = methanol

mer = meridional

MC = metal centered

MIR = mid infrared

MLCT = metal-to-ligand charge transfer

MO = molecular orbital

NADH = nicotinamide-adenine-dinucleotide

NIR = near infrared

NHE = normal hydrogen electrode

OER = one-electron-reduced

OHP = outer Helmholtz Plane

ox = oxidized

P3HT = poly(3-hexylthiophen)

PL = photoluminescence

PVK = polyvinylcarbazol

QRE = quasi reference electrode

red = reduced

RMS = rout mean square

SEM = scanning electron microscopy

TEOA = triethanolamine

TON = turn over number

$TBAPF_6$ = tetrabutylammonium hexafluorophosphate

UV = ultra violet

Vis = visible

WE = working electrode

ZnSe = zinc selenid

List of Figures

1.1 Atmospheric carbon dioxide concentration 10

1.2 The schematic representation of energy as a vector in space and time. 11

1.3 Oil production from 1998 to 2011 12

1.4 The schematic proposal of mimicking photosynthesis 15

1.5 Suggested mechanism for photoinduced charge transfer from a biased organic p-type semiconductor onto a catalyst redox mediator for CO_2 reduction. 18

2.1 Schematic representation of the energy required for the reduction of D to D^- going over an energy intensive transition state. . . . 22

2.2 Schematic representation of the changing surface concentration of the reacting species after the electron transfer is started. . . . 23

2.3 (a) Schematic representation of the typical behavior of electrochemical current over time data according to equation 2.9. (b) When the same data is plotted as $1/time^{1/2}$ the data follow a straight line. 24

2.4 Schematic representation of two layers of charge (the double layer) forming at the interface between the electrode and the electrolyte. The potential drop is confined to this double layer region (also called the outer Helmholtz Plane, OHP) in solution. 26

LIST OF FIGURES

2.5 One-compartment cell for cyclic voltammetry experiments . . . 28

2.6 Quartz cuvette setup for cyclic voltammetry experiments 29

2.7 Typical example of cyclic voltammograms of 1-1 in CO_2 saturated electrolyte solution 30

2.8 H-cell setup with separated anode and cathode compartment for controlled potential electrolysis 31

2.9 Typical current density vs. time plot for potentionstatic CO_2-electrolysis experiment of Re(5,5'-bisphenylethynyl-2,2'bipyridyl)-$(CO)_3Cl$ (1-3) 32

2.10 Illustration of the general UV-Vis spectrometer setup consisting of a deuterium (D2) and tungsten light source, monochromator, the beam selector, sample compartment and the detector. 36

2.11 Illustration of the general PL spectrometer setup consisting of a Xenon arc source, Czerny-Turner type gratings, sample compartment and the PTI PMT detector. 37

2.12 Illustration of the general FTIR spectrometer setup consisting of a MIR light source, an interferometer, the sample chamber and the detector. 38

2.13 Illustration of the the gas tight transmission cell with ZnSe windows built for IR difference absorption spectrum measurement in transmission mode 39

2.14 (a) IR difference absorption spectrum in transmission mode, with and without 5 ml headspace sample after 50 minutes electrolysis experiment of compound 1-3 in solution. (b) Foto of the IR transmission gas cell with two ZnSe windows as it is mounted in the IR spectrometer compartment. 40

2.15 IR difference absorption spectra in transmission mode, with different volumes of a N_2 calibration gas mixture 41

LIST OF FIGURES

2.16 Schematic of a ATR-FTIR ZnSe crystal as reflection element . . 42

2.17 One-compartment cell with a ZnSe reflection element covered with a thin sputtered layer of Pt 43

2.18 (a) Picture of the PTFE cell for ATR-FTIR spectroelectrochemistry with tubes for electrolyte in- and outlet and the Pt counter electrode visible in the back of the cell (b) Foto of the ATR-FTIR setup mounted in the FTIR spectrometer with the three electrode setup and the ZnSe ATR crystal. 44

2.19 GC analysis of a headspace sample (200 μl) after a 25 minutes electrolysis experiment with compound 1-3 in a single-compartment cell 45

2.20 Schematic chemical structures of different rhenium compounds . 47

2.21 Schematic chemical structures of three different rhenium compounds 48

3.1 Exemplary illustration of symmetry elements C_4, C_3, C_2, σ_h and σ_d in an octahedral molecular geometry. 50

3.2 Atmospheric carbon dioxide concentration 51

3.3 Schematic illustration of the splitting of the five-fold degenerate metal d^6-orbitals in two fold degenerate e_g and three fold degenerate t_{2g} orbitals within an octahedral ligand field. 53

3.4 Schematic illustration of the CO molecule forming a sigma donor bond with the rhenium metal center d-orbitals and act as a π^*-acceptor, where the π-orbitals of the CO make a back-bonding (back donation) to the metal d-orbitals 53

3.5 Jablonski-Diagramm for rhenium-(I) bipyridine (bpy) complexes 56

4.1 Schematic energy diagram of the lowest-lying excited states of complexes 1-1 to 1-4.[58] 58

LIST OF FIGURES

4.2 UV-Vis absorption spectra of the rheniumcarbonyl-complexes 1-1 to 1-4 in different solvents 59

4.3 UV-visible absorption spectra of complex 1-1 and complex 1-3 . 60

5.1 IR absorption spectra of Re(2,2'-bipyridyl)(CO)$_3$Cl (1-1). Experimentally measured FTIR difference absorption spectra in KBr (black solid line with squares) and corresponding calculated IR absorption spectra by DFT without (red line) and with (blue dashed line) solvent effects taken into account. 62

5.2 Molecular orbital energy levels of Re(2,2'-bipyridyl)(CO)$_3$Cl calculated by DFT for the frontier orbitals including HOMO-LUMO gap. 65

5.3 Visual representation of molecular frontier orbitals of different rhenium compounds studied in the present work. 67

6.1 Schematic representation of a catalyzed and a non catalyzed reaction mechanism with respect to the energy niveau over the reaction coordinate. 69

6.2 Schematic mechanism for the two electron CO$_2$ reduction 71

6.3 Schematic chemical structures of four different rhenium compounds 72

6.4 Cyclic Voltammograms of 1-1 in nitrogen saturated electrolyte solution on the reductive side (black line with circles) and oxidative side (blue line with circles) 73

6.5 Cyclic Voltammograms and UV-vis absorption spectra 74

6.6 Cyclic Voltammograms of 1-1 in nitrogen (black line with squares) and CO$_2$ (red line with circles) saturated electrolyte solution on Pt 75

LIST OF FIGURES

6.7   Cyclic voltammograms of 1-1 in nitrogen (black line with squares) and CO_2 (red line with circles) saturated electrolyte solution on glassy carbon. 76

6.8   Cyclic Voltammograms of 1-2 and 1-4 in nitrogen (black line with squares) and CO_2 (red line with circles) saturated electrolyte solution. 77

6.9   Cyclic Voltammograms of 1.1 compared to 1-3 in nitrogen (black line with squares) and CO_2 (red line with circles) saturated electrolyte solution. 78

6.10 Cyclic voltammograms of 1-1 in CO_2 saturated electrolyte solution with different catalyst concentrations and scan rates 81

6.11 Schematic chemical structures of three different rhenium compounds with bis(arylimino)acenaphthene derivatives (BIAN-R) ligands (2-1) (2-2) and (2-3) for CO_2 reduction. 84

6.12 Cyclic voltammograms of 2-3 in nitrogen saturated electrolyte solution with two different scan rates 85

6.13 Cyclic Voltammograms of 2-3 in nitrogen (black line with squares) and CO_2 (red line with circles) saturated electrolyte solution with and without H_2O added. 86

6.14 Cyclic Voltammograms of 2-2 and 2-1 in nitrogen (black line with squares) and CO_2 (red line with circles) saturated electrolyte solution. 88

6.15 Headspace gas analysis after potentiostatic CO_2-electrolysis experiment of 2-3 by GC and FTIR measurement 89

6.16 Production of CO vs. time plot for potentionstatic CO_2-electrolysis experiment of 2-3 90

6.17 Proposed mechanism for the pyridinium-catalyzed reduction of CO_2 to methanol 92

LIST OF FIGURES

6.18 Schematic chemical structures of the two different catalyst materials pyridine (1) and pyridazine (2) in pristine and protonated form as pyridinium and pyridazinium 93

6.19 Cyclic Voltammograms of pyridinium and pyridazinium in aqueous solution 94

6.20 Cyclic voltammograms of 50 mM pyridinium and pyridazinium in an aqueous solution of 0.5 M KCl at pH 5.3 96

6.21 Dependence of the catalytic peak current under CO_2 saturation on different pyridinium and pyridazinium concentrations 98

6.22 Current-time measurements for the constant potential electrolysis of 50 mM pyridinium and pyridazinium in an aqueous solution of 0.5 M KCl at pH 5.3 99

6.23 GC analysis of the electrolyte solution during constant potential electrolysis for pyridinium and pyridazinium. 101

7.1 Schematic representation of photocatalytic CO_2 reduction with the catalyst material (C) and the donor (D). 105

7.2 Illustration of the photochemical CO_2 reduction experiment with the light source (A), the gas tight reaction cell (B) and the FTIR measurement cell (C). 106

7.3 Schematic chemical structures of three different rhenium compounds investigated for photocatalysis 106

7.4 Comparison of normalized UV-Visible absorption spectra of three different rheniumcarbonyl-complexes in acetonitrile solution. . . 107

7.5 IR difference absorption spectra (transmission mode) of headspace samples after illumination (red line with circles) of a DMF:TEOA (5:1/v:v) solution 108

LIST OF FIGURES

7.6 Schematic representation of electronic transitions between the electronic ground state and the electronic excited state with vertical transitions (1) and not occuring transitions to the relaxation equilibrium state (2). 110

7.7 Schematic representation of electron transfer from A D to A^+ D^- in a parabola potential energy distribution according to the Marcus theory. 111

8.1 Schematic representation of the electrochemical polymerization of pyrrole to polypyrrole uppon oxidation. 114

8.2 Potentiodynamic electropolymerisation and characterization of pure pyrrol over 70 cycles on a Pt foil 115

8.3 Cyclic Voltammograms of a polypyrrole covered Pt electrode and incoporated catalyst compound 1-1 in N_2 (black line with squares) and CO_2 (red line with circles) saturated electrolyte solution. 116

8.4 Schematic representation of the electrochemical polymerization of the monomer compound (1-3) and the resulting possible chemical substructure of the rhenium sites within the polymer film in which X represents a chloride or a substituted ligand from the reaction medium. 118

8.5 Potentiodynamic formation of rhenium catalyst film on Pt from a catalyst monomer solution of 1-3 119

8.6 Cyclic voltammograms of the rhenium catalyst film on Pt in nitrogen saturated acetonitrile solution 120

8.7 Cyclic voltammograms of the rhenium catalyst film on a Pt plate electrode in nitrogen- and CO_2- saturated electrolyte solution . 121

8.8 Cyclic voltammograms of the rhenium catalyst film on a Pt plate electrode (red line with circles) and a solution of the monomer 1-3 in CO_2-saturated electrolyte solution. 122

LIST OF FIGURES

8.9 Current time curve and UV-Vis absorption spectra for the potentiostatic film formation in nitrogen saturated acetonitrile solution 124

8.10 ATR-FTIR difference absorption spectra of a thick rhenium catalyst film on Pt sputtered onto a ZnSe ATR crystal 126

8.11 Schematic representation of (4,4'-dicarboxyl-2,2'-bipyridyl)Re(CO)$_3$Cl (1-2) attached to an electrode by the metal native ZnO layer and the expected, subsequent CO_2 reduction mechanism a to c. . . . 128

8.12 Cyclic voltammograms of (4,4'-dicarboxyl-2,2'-bipyridyl)Re(CO)$_3$Cl immobilized at a zinc oxide layer in aqueous, N_2 purged solution with a KHPO$_4$ buffer at pH 7 129

9.1 Suggested mechanism for photoinduced charge transfer from a biased organic p-type semiconductor onto pyridinium as a catalyst redox mediator for CO_2 reduction. 132

9.2 Schematic representation of the band structure of a semiconductor 133

9.3 Band structure for a n-type semiconductor not in contact with the electrolyte solution (a) and (b) band-bending if the semiconductor is in contact with the electrolyte. 134

9.4 Band structure for a p-type semiconductor not in contact with the electrolyte solution (a) and (b) band-bending if the semiconductor is in contact with the electrolyte. 135

9.5 Schematic representation of the *Mott − Schottky* analysis for a p-type semiconductor. 136

9.6 Schematic representation of (a) electrochemical reduction and oxidation and (b) optical absorption measurement. 136

9.7 Schematic chemical structure of poly(3-hexylthiophene) (P3HT) (b) Cyclic voltammograms of a P3HT and a blend of P3HT:PCBM covered ITO working electrode 137

LIST OF FIGURES

9.8 Cyclic voltammetry of a pyridine and pyridinium electrolyte solution under dark and white light illumination with a 120 nm thick P3HT covered ITO working electrode. 138

9.9 Photoresponce of a P3HT covered ITO electrode at constant potential 140

9.10 Photocurrent plotted as from the photoresponse experiments of a P3HT covered ITO electrode at constant potential 141

9.11 Temperature increase upon white light irradiation on a ITO electrode in a nitrogen-saturated KCl solution 142

9.12 Chemical structure of (2,2'-bipyridyl)Re(CO)$_3$Cl and its corresponding highest occupied molecular orbital (HOMO) and lowest unoccupied molecular orbital (LUMO) level in electron volt (eV) 144

9.13 PVK and (2,2'-bipyridyl)Re(CO)$_3$Cl (1-1) excitation and photoluminescence spectra 145

9.14 PL quenching experiments with PVK as donor polymer and different amounts of (2,2'-bipyridyl)Re(CO)$_3$Cl (1-1) catalyst as quenching material added on glass/ITO substrate 148

9.15 PL spectra with PVK as donor polymer and of (2,2'-bipyridyl)Re(CO)$_3$Cl (1-1) catalyst as quenching material on glass/ITO substrate. 150

9.16 PL quenching experiments and Stern-Volmer plot with a solid, approx. 9nm thick film of PVK on ITO/glass substrate as donor polymer and different concentrations of (2,2'-bipyridyl)Re(CO)$_3$Cl (1-1) catalyst 152

9.17 Schematic representation of three different conducting polymers investigated in this study 154

9.18 Potentiodynamic electropolymerization of 0.1 M thiophene to polythiophene (8-1) in ACN solution 156

LIST OF FIGURES

9.19 Potentiodynamic electropolymerization of 0.056 M 3-hexylthiophene to poly (3-hexylthiophene) (8-2) in propylene carbonate solution   157

9.20 Potentiodynamic electropolymerization of 8-3   159

9.21 Cyclic voltammograms of poly (3-hexylthiophene) (8-2) electropolymerized on a Pt plate electrode and cyclic voltammograms of a thin film of fac-(2,2'-bipyridyl)Re(CO)$_3$Cl functionalized poly (3-hexylthiophene) (8-3)   160

10.1 Summary of the materials investigated within this work and the obtained efficiencies. Faraday efficiency $\eta_{Faraday}$, energy efficiency η_{Energy} and quantum yield Φ_{CO} for CO_2 reduction.   165

10.2 Schematic representation of the differences between a homogeneous and a heterogeneous process for catalytic CO_2 reduction .   167

10.3 Relation between current densities vs. turnover frequencies of a monolayer of catalyst material for the two electron reduction of CO_2 to CO.   171

10.4 Schematic representation of the characteristic potentials necessary for a CO_2 electrolyser system consisting of two electrodes with a membrane to separated anode and cathode compartment.   173

List of Tables

4.1 Summary of photophysical data of rhenium tetracarbonyl diimino complexes 1-1 to 1-4. The abbreviation LL for compound 1-4 is (2,6-bis-octyloxy-4-formyl)phenylethinyl) 59

5.1 Summary of quantum mechanical calculations of different rhenium compounds studied in the present work. 65

6.1 Summary of controlled potential electrolysis experiments shown in Figure 6.23 and the corresponding calculated Faraday efficiencies. 102

LIST OF TABLES

Bibliography

[1] A. Kerr, Richard, "Even Oil Optimists Expect Energy Demand to Outstrip Supply," *Science*, vol. 317, p. 437, 2007.

[2] K. Richardson, W. Steffen, J. Schellnhuber, Hans, J. Alcamo, T. Barker, M. D. Kammen, R. Leemans, D. Liverman, M. Munasinghe, B. Osman-Elasha, N. Stern, and O. Waever, *Synthesis Report Climatechange- Global Risks, Challenges & Decisions*. 2009.

[3] P. Tans and R. Keeling, "Atmospheric CO_2 at Mauna Loa Observatory," 2013.

[4] S. Enthaler, "Carbon dioxide–the hydrogen-storage material of the future?," *ChemSusChem*, vol. 1, pp. 801–4, Jan. 2008.

[5] *Desertec WhiteBook*. Bonn: Protext Verlag, 4th ed., 2009.

[6] N. S. Lewis and D. G. Nocera, "Powering the planet: chemical challenges in solar energy utilization.," *Proceedings of the National Academy of Sciences of the United States of America*, vol. 103, pp. 15729–35, Oct. 2006.

[7] M. Fischer, M. Werber, and P. V. Schwartz, "Batteries: Higher energy density than gasoline?," *Energy Policy*, vol. 37, pp. 2639–2641, July 2009.

[8] L. R. Raymond, A. Gould, J. J. Hamre, D. J. O'Reilly, and D. H. Yergin, "Hardtruths Facing the Hard Truths about Energy," tech. rep., National Petroleum Council, 2007.

[9] T. Appenzeller, "End of Cheap Oil," *National Geographic*, vol. 205, no. 6, p. 80, 2004.

[10] J. Murray and D. King, "Oil's tipping point has passed," *Nature*, vol. 481, pp. 433 – 435–8, 2012.

[11] IHS/CERA, "Cambridge Energy Research Associates Finding the Critical Numbers: What Are the Real Decline Rates for Global Oil Production?," 2007.

[12] IEA, "International Energy Agency World Energy Outlook 2008," tech. rep., 2008.

[13] DOE/EIA, "Annual Energy Outlook," tech. rep., US Energy Information Administration, 2011.

[14] D. H. Apaydin, E. D. Głowacki, E. Portenkirchner, and N. S. Sariciftci, "Direct electrochemical capture and release of carbon dioxide using an industrial organic pigment: quinacridone.," *Angewandte Chemie (International ed. in English)*, vol. 53, pp. 6819–22, June 2014.

[15] S. Stucki, "Coupled CO_2 recovery from the atmosphere and water electrolysis: Feasibility of a new process for hydrogen storage," *International Journal of Hydrogen Energy*, vol. 20, pp. 653–663, Aug. 1995.

[16] M. Specht, F. Staiss, A. Bandi, and T. Weimer, "Comparison of the renewable transportation fuels, liquid hydrogen and methanol, with gasoline - Energetic and economic aspects," *International Journal of Hydrogen Energy*, vol. 23, pp. 387–396, May 1998.

[17] G. A. Olah, A. Goeppert, and G. K. S. Prakash, "Chemical recycling of carbon dioxide to methanol and dimethyl ether: from greenhouse gas to renewable, environmentally carbon neutral fuels and synthetic hydrocarbons.," *The Journal of organic chemistry*, vol. 74, pp. 487–98, Jan. 2009.

[18] T. Abe and M. Kaneko, "Reduction catalysis by metal complexes confined in a polymer matrix," *Progress in Polymer Science*, vol. 28, pp. 1441–1488, Oct. 2003.

[19] B. Kumar, M. Llorente, J. Froehlich, T. Dang, A. Sathrum, and C. P. Kubiak, "Photochemical and photoelectrochemical reduction of CO_2.," *Annual review of physical chemistry*, vol. 63, pp. 541–69, Jan. 2012.

[20] E. E. Benson, C. P. Kubiak, A. J. Sathrum, and J. M. Smieja, "Electrocatalytic and homogeneous approaches to conversion of CO_2 to liquid fuels.," *Chemical Society reviews*, vol. 38, pp. 89–99, Jan. 2009.

[21] S. C. Roy, O. K. Varghese, M. Paulose, and C. A. Grimes, "Toward solar fuels: photocatalytic conversion of carbon dioxide to hydrocarbons.," *ACS nano*, vol. 4, pp. 1259–78, Mar. 2010.

[22] M. V. V. S. Reddy, K. V. Lingam, and T. K. G. Rao, "Molecular Physics : An International Journal at the Interface Between Chemistry and Physics Studies of radicals in oxalate systems," *Molecular Physics*, vol. 41, no. 6, pp. 1493–1500, 1980.

[23] H. A. Schwarz, C. Creutz, and N. Sutin, "Cobalt(1) Polypyridine Complexes. Redox and Substitutional Kinetics and Thermodynamics in the Aqueous 2,2-Bipyridine and 4,4-Dimethyl-2,2- bipyridine Series Studied by the Pulse-Radiolysis Technique," *Journal of Amercian Chemical Society*, vol. 24, no. 1, pp. 433–439, 1985.

[24] H. A. Schwarz and W. Dodson, "Reduction Potentials of C02- and the Alcohol Radicals," *Journal of Physical Chemistry*, vol. 93, pp. 409–414, 1989.

[25] C. Delacourt, P. L. Ridgway, J. B. Kerr, and J. Newman, "Design of an Electrochemical Cell Making Syngas (CO + H_2) from CO_2 and H_2O Reduction at Room Temperature," *Journal of The Electrochemical Society*, vol. 155, no. 1, p. B42, 2008.

[26] G. Knör and U. W. E. Monkowius, "Photosensitization and photocatalysis in bioinorganic, bio-organometallic and biomimetic systems," *Advanced Inorganic Chemistry*, vol. 63, pp. 235–289, 2011.

[27] J. J. Concepcion, R. L. House, J. M. Papanikolas, and T. J. Meyer, "Chemical approaches to artificial photosynthesis.," *Proceedings of the National Academy of Sciences of the United States of America*, vol. 109, pp. 15560–4, Sept. 2012.

[28] Y. Lu, Z.-y. Jiang, S.-w. Xu, and H. Wu, "Efficient conversion of CO_2 to formic acid by formate dehydrogenase immobilized in a novel alginate–silica hybrid gel," *Catalysis Today*, vol. 115, pp. 263–268, June 2006.

[29] B. El-zahab, D. Donnelly, and P. Wang, "Particle-Tethered NADH for Production of Methanol From CO_2 Catalyzed by Coimmobilized Enzymes," *Biotechnology and bioengineering*, vol. 99, no. 3, pp. 508–514, 2008.

[30] Q. Sun, Y. Jiang, Z. Jiang, L. Zhang, X. Sun, and J. Li, "Green and Efficient Conversion of CO 2 to Methanol by Biomimetic Coimmobilization of Three Dehydrogenases in Protamine-Templated Titania," *Industrial & Engineering Chemistry Research*, vol. 48, pp. 4210–4215, May 2009.

[31] E. B. Cole, P. S. Lakkaraju, D. M. Rampulla, A. J. Morris, E. Abelev, and A. B. Bocarsly, "Using a one-electron shuttle for the multielectron reduction of CO_2 to methanol: kinetic, mechanistic, and structural insights.," *Journal of the American Chemical Society*, vol. 132, pp. 11539–51, Aug. 2010.

[32] B. Kumar, J. M. Smieja, A. F. Sasayama, and C. P. Kubiak, "Tunable, light-assisted co-generation of CO and H2 from CO_2 and H_2O by Re(bipy-tbu)$(CO)_3$Cl and p-Si in non-aqueous medium.," *Chemical communications (Cambridge, England)*, vol. 48, pp. 272–4, Jan. 2012.

[33] E. E. Barton, D. M. Rampulla, and A. B. Bocarsly, "Selective solar-driven reduction of CO_2 to methanol using a catalyzed p-GaP based photoelectrochemical cell.," *Journal of the American Chemical Society*, vol. 130, pp. 6342–4, May 2008.

[34] H. Gerischer, "Electron-transfer kinetics of redox reactions at the semiconductor/electrolyte contact. A new approach," *The Journal of Physical Chemistry*, vol. 95, pp. 1356–1359, Feb. 1991.

[35] A. J. Heeger, N. S. Sariciftci, and E. B. Namdas, *Semiconducting and Metallic Polymers*. New York: Oxford University Press, 2010.

[36] C. Rüssel and W. Jaenicke, "Rate constants, activation free enthalpies and activation entropies of the electrochemical reduction of 1,4-diazines in DMF," *Electrochimica Acta*, vol. 27, pp. 1745–1750, Dec. 1982.

[37] a. M. Bond, T. L. E. Henderson, D. R. Mann, T. F. Mann, W. Thormann, and C. G. Zoski, "A fast electron transfer rate for the oxidation of ferrocene in acetonitrile or dichloromethane at platinum disk ultramicroelectrodes," *Analytical Chemistry*, vol. 60, pp. 1878–1882, Sept. 1988.

[38] C. H. Hamann, A. Hamnett, and W. Vielstich, *Electrochemistry*. Weinheim: WILEY-VCH, 1998.

[39] A. J. Bard and L. R. Faulkner, *Electrochemical Methods: Fundamentals and Applications*. WILEY-VCH, 2 edition ed., 2000.

[40] C. H. Hamann, A. Hamnett, and W. Vielstich, *Electrochemistry.* Weinheim: WILEY-VCH, 2nd ed., 2007.

[41] E. Fujita, D. J. Szalda, C. Creutz, B. National, and C. O. Chemistry, "Carbon Dioxide Activation: Thermodynamics of CO, Binding and the Involvement of Two Cobalt Centers in the Reduction of CO, by a Cobalt(1) Macrocycle," *Journal of Amercian Chemical Society*, vol. 110, no. 1, pp. 4870–4871, 1988.

[42] C. M. Cardona, W. Li, A. E. Kaifer, D. Stockdale, and G. C. Bazan, "Electrochemical considerations for determining absolute frontier orbital energy levels of conjugated polymers for solar cell applications.," *Advanced materials (Deerfield Beach, Fla.)*, vol. 23, pp. 2367–71, May 2011.

[43] E. Portenkirchner, K. Oppelt, C. Ulbricht, D. a.M. Egbe, H. Neugebauer, G. Knör, and N. S. Sariftci, "Electrocatalytic and photocatalytic reduction of carbon dioxide to carbon monoxide using the alkynyl-substituted rhenium(I) complex(5,5-bisphenylethynyl-2,2-bipyridyl)Re(CO)3Cl," *Journal of Organometallic Chemistry*, vol. 716, pp. 19–25, Oct. 2012.

[44] Z. K. Lopez-Castillo, S. N. V. K. Aki, M. A. Stadtherr, and J. F. Brennecke, "Enhanced Solubility of Oxygen and Carbon Monoxide in CO_2-Expanded Liquids," *Ind. Eng. Chem. Res.*, vol. 45, pp. 5351–5360, 2006.

[45] S. Cosnier, A. Deronzier, and J.-C. Moutet, "Electrochemical coating of a platinum electrodeby a polypyrrole film containing the fac-Re(bipyridine)(CO)$_3$Cl system," *Journal of Electroanalytical Chemistry*, vol. 207, pp. 315–321, 1986.

[46] J. M. Smieja and C. P. Kubiak, "Re(bipy-tBu)(CO)3Cl-improved catalytic activity for reduction of carbon dioxide: IR-spectroelectrochemical and mechanistic studies.," *Inorganic chemistry*, vol. 49, pp. 9283–9, Oct. 2010.

[47] J. Hawecker, J.-m. Lehn, and R. Ziessel, "Photochemical and Electrochemical Reduction of Carbon Dioxide to Carbon Monoxide Mediated by (2,2'-Bipyridine) tricarbonylchlororhenium (I) and Related Complexes as Homogeneous Catalysts," *Helvetica Chimica Acta*, vol. 69, pp. 1990–2012, 1986.

[48] D. C. Harris and M. D. Bertolucci, *Symmetry and Spectroscopy: An Introduction to Vibrational and Electronic Spectroscopy*. New York: Dover Publishing, 1978.

[49] E. Portenkirchner, N. S. Sariçiftçi, K. Oppelt, and D. A. M. Knör, GüntherEgbe, "Electro- and photo-chemistry of rhenium and rhodium complexes for carbon dioxide and proton reduction: a mini review," *Nanomaterials and Energy*, vol. 2, pp. 134–147, May 2013.

[50] P. Atkins and J. de Paula, *Atkins Physical Chemistry*. Oxford New York: Oxford University Press, 8th ed., 2006.

[51] E. Riedel, *Anorganische Chemie*. Berlin, New York: De Gruyter, 2nd ed., 1990.

[52] G. M. Barrow and G. W. Herzog, *Physikalische Chemie*. Wien: Bohmann-Verlag, 6 ed., 1984.

[53] H. Takeda, K. Koike, and T. Morimoto, "Photochemistry and photocatalysis of rhenium (I) diimine complexes," *Inorganic Photochemistry*, vol. 63, no. I, pp. 137–186, 2011.

[54] G. Knör*, M. Leirer, and A. Vogler*, "Synthesis, characterization and spectroscopic properties of 1,2-diiminetricarbonylrhenium(I)chloride complexes with o-benzoquinone diimines as ligands," *Journal of Organometallic Chemistry*, vol. 610, pp. 16–19, Sept. 2000.

[55] U. Monkowius, S. Ritter, B. König, M. Zabel, and H. Yersin, "Synthesis, Characterisation and Ligand Properties of Novel Bi-1,2,3-triazole Ligands," *European Journal of Inorganic Chemistry*, vol. 2007, pp. 4597–4606, Oct. 2007.

[56] M. Leirer, G. Knör, and A. Vogler, "Electronic spectra of 1,2-diiminetricarbonylrhenium(I)chloride complexes with imidazole derivatives as ligands," *Inorganica Chimica Acta*, vol. 288, pp. 150–153, May 1999.

[57] A. Beer, "Bestimmung der Absorption des roten Lichts in farbigen Flüssigkeiten," *Annalen der Physik und Chemie*, vol. Bd. 86, pp. 78–88, 1852.

BIBLIOGRAPHY

[58] H. Takeda, K. Koike, and T. Morimoto, "Photochemistry and photocatalysis of rhenium (I) diimine complexes," *Inorganic Photochemistry*, vol. 63, no. I, pp. 137–186, 2011.

[59] B. P. Sullivan, C. M. Bolinger, D. Conrad, W. J. Vining, and T. J. Meyer, "One- and two-electron pathways in the electrocatalytic reduction of CO_2 by fac-Re(bpy)(CO)$_3$Cl (bpy = 2,2-bipyridine)," *Journal of the Chemical Society, Chemical Communications*, vol. 20, no. 20, p. 1414, 1985.

[60] K. Koike, H. Hori, M. Ishizuka, J. R. Westwell, K. Takeuchi, T. Ibusuki, K. Enjouji, H. Konno, K. Sakamoto, and O. Ishitani, "Key Process of the Photocatalytic Reduction of CO_2 Using [Re(4,4'-X 2-bipyridine)(CO)$_3$PR$_3$] + (X=CH$_3$, H, CF$_3$; PR$_3$ = Phosphorus Ligands): Dark Reaction of the One-Electron-Reduced Complexes with CO_2," *Organometallics*, vol. 16, pp. 5724–5729, Dec. 1997.

[61] W. Kaim, A. Klein, and T. Scheiring, "EPR study of paramagnetic rhenium(I) complexes (bpy)Re(CO)3X relevant to the mechanism of electrocatalytic CO_2 reduction.," *Journal of the Chemical Society, Perkin Transactions*, pp. 2569–2572, 1997.

[62] K. Oppelt, D. a.M. Egbe, U. Monkowius, M. List, M. Zabel, N. S. Sariciftci, and G. Knör, "Luminescence and spectroscopic studies of organometallic rhodium and rhenium multichromophore systems carrying polypyridyl acceptor sites and phenylethynyl antenna subunits," *Journal of Organometallic Chemistry*, vol. 696, pp. 2252–2258, May 2011.

[63] K. A. Walters, L. Premvardhan, Y. Liu, L. Peteanu, and K. S. Schanze, "Metal-to-ligand charge transfer absorption in a rhenium (I) complex that contains a -conjugated bipyridine acceptor ligand.," *Chemical Physics Letters*, no. 339, pp. 255–262, 2001.

[64] Y. Liu, Y. Li, and K. S. Schanze, "Photophysics of π-conjugated oligomers and polymers that contain transition metal complexes," *Journal of Photochemistry and Photobiology C: Photochemistry Reviews*, vol. 3, pp. 1–23, June 2002.

[65] M. Leirer, G. Knör, and A. Vogler, "Electronic spectra of 1,2-diiminetricarbonylrhenium(I)chloride complexes with imidazole derivatives as ligands," *Inorganica Chimica Acta*, vol. 288, pp. 150–153, May 1999.

[66] W. Greiner, *Quantum Mechanics - An Introduction.* Berlin: Springer-Verlag Berlin Heidelberg New York, 4th editio ed., 2001.

[67] A. Szabo and N. S. Ostlund, *Modern Quantum Chemistry: Introduction to Advanced Electronic Structure Theory.* New York: Dover Pubn Inc, new editio ed., 1996.

[68] M. J. Frisch, G. W. Trucks, H. B. Schlegel, G. E. Scuseria, M. A. Robb, J. R. Cheeseman, G. Scalmani, V. Barone, B. Mennucci, G. A. Petersson, H. Nakatsuji, M. Caricato, X. Li, H. P. Hratchian, A. F. Izmaylov, J. Bloino, G. Zheng, J. L. Sonnenberg, M. Hada, M. Ehara, K. Toyota, R. Fukuda, J. Hasegawa, M. Ishida, T. Nakajima, Y. Honda, O. Kitao, H. Nakai, T. Vreven, J. A. Montgomery, Jr., J. E. Peralta, F. Ogliaro, M. Bearpark, J. J. Heyd, E. Brothers, K. N. Kudin, V. N. Staroverov, R. Kobayashi, J. Normand, K. Raghavachari, A. Rendell, J. C. Burant, S. S. Iyengar, J. Tomasi, M. Cossi, N. Rega, J. M. Millam, M. Klene, J. E. Knox, J. B. Cross, V. Bakken, C. Adamo, J. Jaramillo, R. Gomperts, R. E. Stratmann, O. Yazyev, A. J. Austin, R. Cammi, C. Pomelli, J. W. Ochterski, R. L. Martin, K. Morokuma, V. G. Zakrzewski, G. A. Voth, P. Salvador, J. J. Dannenberg, S. Dapprich, A. D. Daniels, . Farkas, J. B. Foresman, J. V. Ortiz, J. Cioslowski, and D. J. Fox, "Gaussian 09 Revision D.01." Gaussian Inc. Wallingford CT 2009.

[69] A. D. Becke, "Density-functional exchange-energy approximation with correct asymptotic behavior," *Physical Review A,* vol. 38, pp. 3098–3100, Sept. 1988.

[70] C. Lee, W. Yang, and R. G. Parr, "Development of the Colle-Salvetti correlation-energy formula into a functional of the electron density," *Physical Review B,* vol. 37, pp. 785–789, Jan. 1988.

[71] A. D. Becke, "Density-functional thermochemistry. III. The role of exact exchange," *The Journal of Chemical Physics,* vol. 98, no. 7, p. 5648, 1993.

[72] W. J. Hehre, "Self-Consistent Molecular Orbital Methods. XIV. An Extended Gaussian-Type Basis for Molecular Orbital Studies of Organic Molecules. Inclusion of Second Row Elements," *The Journal of Chemical Physics,* vol. 56, no. 11, p. 5255, 1972.

[73] W. J. Hehre, "Self—Consistent Molecular Orbital Methods. XII. Further Extensions of Gaussian—Type Basis Sets for Use in Molecular Orbital

Studies of Organic Molecules," *The Journal of Chemical Physics*, vol. 56, no. 5, p. 2257, 1972.

[74] V. a. Rassolov, M. a. Ratner, J. a. Pople, P. C. Redfern, and L. a. Curtiss, "6-31G* basis set for third-row atoms," *Journal of Computational Chemistry*, vol. 22, pp. 976–984, July 2001.

[75] P. J. Hay and W. R. Wadt, "Ab initio effective core potentials for molecular calculations. Potentials for the transition metal atoms Sc to Hg," *The Journal of Chemical Physics*, vol. 82, no. 1, p. 270, 1985.

[76] W. R. Wadt and P. J. Hay, "Ab initio effective core potentials for molecular calculations. Potentials for main group elements Na to Bi," *The Journal of Chemical Physics*, vol. 82, no. 1, p. 284, 1985.

[77] P. J. Hay and W. R. Wadt, "Ab initio effective core potentials for molecular calculations. Potentials for K to Au including the outermost core orbitals," *The Journal of Chemical Physics*, vol. 82, no. 1, p. 299, 1985.

[78] C. M. Rohlfing, P. J. Hay, and R. L. Martin, "An effective core potential investigation of Ni, Pd, and Pt and their monohydrides," *The Journal of Chemical Physics*, vol. 85, no. 3, p. 1447, 1986.

[79] A. Gilbert, *Essentials of molecular photochemistry*. Cambridge: Blackwell Scientific Publications, 1991.

[80] L. G. Wade, *Organic Chemistry*. New Jersey: Englewood Cliffs, N.J. : Prentice-Hall, 1987.

[81] A. P. Scott and L. Radom, "Harmonic Vibrational Frequencies: An Evaluation of Hartree - Fock, Moller- Plesset, Quadratic Configuration Interaction, Density Functional Theory, and Semiempirical Scale Factors," *The Journal of Physical Chemistry*, vol. 100, pp. 16502–16513, Jan. 1996.

[82] G. A. Olah, "Beyond oil and gas: the methanol economy.," *Angewandte Chemie (International ed. in English)*, vol. 44, pp. 2636–9, Apr. 2005.

[83] T. R. O. Toole, B. P. Sullivan, M. R. Bruce, L. D. Margerum, R. W. Murray, and T. J. Meyer, "Electrocatalytic reduction of CO_2, by a complex of rhenium in thin polymeric films," *Journal of Electroanalytical Chemistry*, vol. 259, pp. 217–239, 1989.

[84] F. M. Romero and R. Ziessel, "A Straightforward Synthesis of 5-Bromo and 5,5'-Dibromo-2,2'-Bipyridines," *Tetrahedron Letters*, vol. 36, no. 36, pp. 6471–6474, 1995.

[85] K. Sanechika, T. Yamamoto, and A. Yamamoto, "Palladium Catalyzed C-C Coupling for Synthesis of pi-Conjugated Polymers Composed of Arylene and Ethynylene Units," *Bulletin of the Chemical Society of Japan*, vol. 57, no. 3, pp. 752–755, 1984.

[86] U. Monkowius, S. Ritter, B. König, M. Zabel, and H. Yersin, "Synthesis, Characterisation and Ligand Properties of Novel Bi-1,2,3-triazole Ligands," *European Journal of Inorganic Chemistry*, vol. 2007, pp. 4597–4606, Oct. 2007.

[87] U. Monkowius, Y. Svartsov, T. Fischer, M. Zabel, and H. Yersin, "Synthesis, crystal structures, and electronic spectra of (1,8-naphthyridine)ReI(CO)3Cl and [(1,8-naphthyridine)CuI(DPEPhos)]PF6," *Inorganic Chemistry Communications*, vol. 10, pp. 1473–1477, Dec. 2007.

[88] J. Hawecker, J.-m. Lehn, and R. Ziessel, "Electrocatalytic Reduction of Carbon Dioxide Mediated by Re(bipy)(CO)3Cl (bipy = 2,2-bipyridine)," *Journal of Chemical Society, Chemical Communication*, vol. 6, pp. 328–330, 1984.

[89] J. Hawecker, J.-m. Lehn, and R. Ziessel, "Photochemical and Electrochemical Reduction of Carbon Dioxide to Carbon Monoxide Mediated by (2 , 2 -Bipyridine) tricarbonylchlororhenium (I) and Related Complexes as Homogeneous Catalysts)," *Helvetica Chimica Acta*, vol. 69, pp. 1990–2012, 1986.

[90] F. P. A. Johnson, M. W. George, F. Hartl, and J. J. Turner, "Electrocatalytic Reduction of CO_2 Using the Complexes [Re(bpy)(CO)$_3$ L]n (n=+1, L= P(OEt)$_3$, CH_3CN; n=0, L=Cl$^-$, Otf; bpy = 2,2'-Bipyridine; Otf=CF_3SO_3) as Catalyst Precursors: Infrared Spectroelectrochemical Investigation," *Organometallics*, vol. 15, pp. 3374–3387, Jan. 1996.

[91] C. Amatore, J. Pinson, and J.-M. Savéant, "Trace crossings in cyclic voltammetry and electrochemic electrochemical inducement of chemical reactions," *Journal of Electroanalytical Chemistry*, vol. 107, pp. 59–74, 1980.

[92] T. F. Connors, J. V. Arena, and J. F. Rusling, "Electrocatalytic Reduction of Viclnal Dibromides by Vitamin B12," *Journal of Physical Chemistry*, vol. 92, no. 10, pp. 2810–2816, 1988.

[93] A. Houmam, E. M. Hamed, and I. W. J. Still, "A Unique Autocatalytic Process and Evidence for a Concerted-Stepwise Mechanism Transition in the Dissociative Electron-Transfer Reduction of Aryl Thiocyanates," *Techniques*, vol. 185, no. 7, pp. 7258–7265, 2003.

[94] P. Christensen, A. Hamnett, S. A. V. G. Muir, J. A. Timneyb, S. Section, N. T. College, and W. Ne, "An In Situ Infrared Study of CO, Reduction catalysed by Rhenium Tricarbonyl Bipyridyl Derivatives," *Journal of Chemical Society, Dalton Transactions*, vol. 9, pp. 1455–1463, 1992.

[95] J. M. Saveant and E. Vianello, "Potential-sweep chronoamperometry theory of kinetic currents in the case of a first order chemical reactions preceding the electron-transfer process," *Electrochimica Acta*, vol. 8, pp. 905–923, 1963.

[96] D. L. Dubois, A. Miedanerlt, and R. C. Haltiwangert, "Electrochemical Reduction of CO_2 Catalyzed by [Pd(triphosphine)(solvent)]$(BF_4)_2$ Complexes:Synthetic and Mechanistic Studies," *Journal of Amercian Chemical Society*, vol. 113, pp. 8753–8764, 1991.

[97] P. W. Atkins, *Physikalische Chemie*. WILEY-VCH, 3rd ed., 2001.

[98] T. Kern, U. Monkowius, M. Zabel, and G. Knör, "Mononuclear Copper(I) Complexes Containing Redox-Active 1,2-Bis(aryl-imino)acenaphthene Acceptor Ligands: Synthesis, Crystal Structures and Tuneable Electronic Properties," *European Journal of Inorganic Chemistry*, vol. 2010, pp. 4148–4156, Sept. 2010.

[99] G. Knör, M. Leirer, T. E. Keyes, J. G. Vos, and A. Vogler, "Non-Luminescent 1,2-Diiminetricarbonylrhenium (I) Chloride Complexes – Synthesis , Electrochemical and Spectroscopic Properties of Re(DIAN)(CO)$_3$Cl with DIAN p -Substituted Bis(arylimino)acenaphthene," *European Journal of Inorganic Chemistry*, no. I, pp. 749–751, 2000.

[100] C. Topf, U. Monkowius, and G. Knör, "Design, synthesis and characterization of a modular bridging ligand platform for bio-inspired hydrogen

production," *Inorganic Chemistry Communications*, vol. 21, pp. 147–150, July 2012.

[101] E. Portenkirchner, E. Kianfar, N. S. Sariciftci, and G. Knör, "Two-electron carbon dioxide reduction catalyzed by rhenium(I) bis(imino)acenaphthene carbonyl complexes.," *ChemSusChem*, vol. 7, pp. 1347–51, May 2014.

[102] G. Seshadri, C. Lin, and A. B. Bocarsly, "A new homogeneous electrocatalyst to methanol at low overpotential for the reduction of carbon dioxide," *Journal of Electroanalytical Chemistry*, vol. 372, pp. 145–150, 1994.

[103] A. J. Morris, R. T. McGibbon, and A. B. Bocarsly, "Electrocatalytic carbon dioxide activation: the rate-determining step of pyridinium-catalyzed CO_2 reduction.," *ChemSusChem*, vol. 4, pp. 191–6, Feb. 2011.

[104] E. E. Barton, D. M. Rampulla, and A. B. Bocarsly, "Selective solar-driven reduction of CO_2 to methanol using a catalyzed p-GaP based photoelectrochemical cell.," *Journal of the American Chemical Society*, vol. 130, pp. 6342–4, May 2008.

[105] A. B. Bocarsly, Q. D. Gibson, A. J. Morris, R. P. L'Esperance, Z. M. Detweiler, P. S. Lakkaraju, E. L. Zeitler, and T. W. Shaw, "Comparative Study of Imidazole and Pyridine Catalyzed Reduction of Carbon Dioxide at Illuminated Iron Pyrite Electrodes," *ACS Catalysis*, vol. 2, pp. 1684–1692, Aug. 2012.

[106] J. a. Keith and E. a. Carter, "Theoretical insights into pyridinium-based photoelectrocatalytic reduction of CO_2.," *Journal of the American Chemical Society*, vol. 134, pp. 7580–3, May 2012.

[107] C. Costentin, J. C. Canales, B. Haddou, and J.-M. Savéant, "Electrochemistry of acids on platinum. Application to the reduction of carbon dioxide in the presence of pyridinium ion in water.," *Journal of the American Chemical Society*, vol. 135, pp. 17671–4, Nov. 2013.

[108] F. Nachod and E. A. Braude, *Determination of Organic Structures by Physical Methods*. New York: Academic Press, 1955.

[109] C.-H. Lim, A. M. Holder, and C. B. Musgrave, "Mechanism of homogeneous reduction of CO2 by pyridine: proton relay in aqueous solvent

and aromatic stabilization.," *Journal of the American Chemical Society*, vol. 135, pp. 142–54, Jan. 2013.

[110] E. Portenkirchner, C. Enengl, S. Enengl, G. Hinterberger, S. Stefanie, D. Apaydin, H. Neugebauer, G. Knör, and S. Sariciftci, "A Comparison of Pyridazine and Pyridine as Electrocatalysts for the Reduction of Carbon Dioxide to Methanol," *ChemElectroChem*, 2014.

[111] H. Takeda, K. Koike, H. Inoue, and O. Ishitani, "Development of an efficient photocatalytic system for CO_2 reduction using rhenium(I) complexes based on mechanistic studies.," *Journal of the American Chemical Society*, vol. 130, pp. 2023–31, Feb. 2008.

[112] T. Yui, Y. Tamaki, K. Sekizawa, and O. Ishitani, "Photocatalytic reduction of CO: from molecules to semiconductors.," *Topics in current chemistry*, vol. 303, pp. 151–84, Jan. 2011.

[113] H. Takeda and O. Ishitani, "Development of efficient photocatalytic systems for CO_2 reduction using mononuclear and multinuclear metal complexes based on mechanistic studies," *Coordination Chemistry Reviews*, vol. 254, pp. 346–354, Feb. 2010.

[114] S. Kozuch and J. M. L. Martin, ""Turning Over" Definitions in Catalytic Cycles," *ACS Catalysis*, vol. 2, pp. 2787–2794, Dec. 2012.

[115] T. Yoshida, T. Iida, T. Shirasagi, R.-J. Lin, and M. Kaneko, "Electrocatalytic reduction of carbon dioxide in aqueous medium by bis(2,2:6,2-terpyridine)cobalt(II) complex incorporated into a coated polymer membrane," *Journal of Electroanalytical Chemistry*, vol. 344, pp. 355–362, Jan. 1993.

[116] E. Portenkirchner, S. Schlager, D. Apaydin, K. Oppelt, M. Himmelsbach, D. a. M. Egbe, H. Neugebauer, G. Knör, T. Yoshida, and N. S. Sariciftci, "Using the Alkynyl-Substituted Rhenium(I) Complex (4,4'-Bisphenyl-Ethynyl-2,2'-Bipyridyl)Re(CO)3Cl as Catalyst for CO2 Reduction—Synthesis, Characterization, and Application," *Electrocatalysis*, Oct. 2014.

[117] R. a. Marcus, "On the Theory of Oxidation-Reduction Reactions Involving Electron Transfer. I," *The Journal of Chemical Physics*, vol. 24, no. 5, p. 966, 1956.

[118] S. Arrhenius, "Über die Dissociationswärme und den Einflußder Temperatur auf den Dissociationsgrad der Elektrolyte," *Zeitschrift für Physikalische Chemie*, vol. 4, pp. 96–116, 1889.

[119] S. Shukla, S. B. Halligudi, and M. M. Taqui Khan, "Reduction of CO_2 by molecular hydrogen to formic acid and formaldehyde and their decomposition to CO and H_2O," *Journal of Molecular Catalysis*, vol. 57, pp. 47–60, 1989.

[120] T. Yoshida, K. Kamato, M. Tsukamoto, T. Iida, D. Schlettwein, D. Wöhrle, and M. Kaneko, "Selective electroacatalysis for CO_2 reduction in the aqueous phase using cobalt phthalocyanine/poly-4-vinylpyridine modified electrodes," *Journal of Electroanalytical Chemistry*, vol. 385, pp. 209–225, Apr. 1995.

[121] T. Yoshida, K. Tsutsumida, S. Teratani, K. Yasufuku, and M. Kaneko, "Electrocatalytic Reduction of CO_2 in Water by [Re(bpy)(CO)$_3$Br] and [Re(terpy)(CO)$_3$Br] Complexes incorporated into Coated Nafion Membrane (bpy=2,2-bipyridine; terpy=2,2:6,2-terpyridine)," *Journal of Chemical Society, Chemical Communication*, pp. 631–633, 1993.

[122] A. Zhang, W. Zhang, J. Lu, G. G. Wallace, and J. Chen, "Electrocatalytic Reduction of Carbon Dioxide by Cobalt-Phthalocyanine-Incorporated Polypyrrole," *Electrochemical and Solid-State Letters*, vol. 12, no. 8, p. E17, 2009.

[123] T. R. O. Toole, L. D. Margerum, T. D. Westmoreland, W. J. Vining, R. W. Murray, and T. J. Meyer, "Electrocatalytic Reduction of CO_2 at a Chemically Modified Electrode," *Journal of Chemical Society, Chemical Communication*, pp. 1416–1417, 1985.

[124] N. N. Glagolev, V. M. Misin, N. L. Zaichenko, V. N. Khandozhko, N. Y. Kolobova, and M. I. Cherkashin, "Polymerization of diphenyldiacetylene and of tolan in the presence of carbonyl cobalt complexes, structure and properties of the polymer," *Polymer Science U.S.S.R.*, vol. 28, no. 10, pp. 2359–2366, 1986.

[125] J. Heinze, "Cyclovoltammetrie - die ,,Spektroskopie" des Elektrochemikers," *Angewandte Chemie*, vol. 11, pp. 823 – 916, 1984.

[126] I. Breikss and D. Abruna, "Electrochemical and mechanistic studies of Re(CO)(dmbpy)Cl and their relation to the catalytic reduction of CO_2," *Journal of Electroanalytical Chemistry*, vol. 201, pp. 347–358, 1986.

[127] F. Cecchet, M. Alebbi, C. A. Bignozzi, and F. Paolucci, "Efficiency enhancement of the electrocatalytic reduction of CO_2: fac-[Re(v-bpy)(CO)3Cl] electropolymerized onto mesoporous TiO2 electrodes," *Inorganica Chimica Acta*, vol. 359, pp. 3871–3874, Sept. 2006.

[128] C. Pouchert, *The Aldrich Library of FT-IR Spectra*. Wiley, 1997.

[129] M. Pohjakallio, G. Sundholm, and P. Talonen, "In situ FTIR studies on the redox properties of poly (thiophene-3-methanol) in PF 6 and C10 4 electrolytes," *Journal of Electroanalytical Chemistry*, vol. 406, pp. 165–174, 1996.

[130] J. Garcia-Cañadas, A. Lafuente, G. Rodriiguez, M. L. Marcos, and J. G. Velasco, "Mechanism of electropolymerization of phenylacetylene in propylene carbonate," *Journal of Electroanalytical Chemistry*, vol. 565, pp. 57–64, Apr. 2004.

[131] E. Portenkirchner, J. Gasiorowski, K. Oppelt, S. Schlager, C. Schwarzinger, H. Neugebauer, G. Knör, and N. S. Sariciftci, "Electrocatalytic Reduction of Carbon Dioxide to Carbon Monoxide by a Polymerized Film of an Alkynyl-Substituted Rhenium(I) Complex.," *ChemCatChem*, vol. 5, pp. 1790–1796, July 2013.

[132] C. a. Koval and J. N. Howard, "Electron transfer at semiconductor electrode-liquid electrolyte interfaces," *Chemical Reviews*, vol. 92, pp. 411–433, May 1992.

[133] A. Bott, "Electrochemistry of Semiconductors," *Current Separations*, vol. 3, pp. 87–91, 1998.

[134] S. M. Sze and K. N. Kwok, *Physics of Semiconductor Devices*. WILEY-VCH, 3rd ed., 2007.

[135] N. Sariciftci and A. Heeger, "Photophysics of semiconducting polymer-C60 composites: A comparative study," *Synthetic Metals*, vol. 70, pp. 1349–1352, Mar. 1995.

[136] P. Gründler, A. Kirbs, and L. Dunsch, "Modern thermoelectrochemistry.," *Chemphyschem : a European journal of chemical physics and physical chemistry*, vol. 10, pp. 1722–46, Aug. 2009.

[137] P. Gründler and A. Beckmann, "Hydrodynamics with heated microelectrodes.," *Analytical and bioanalytical chemistry*, vol. 379, pp. 261–5, May 2004.

[138] S. Wang, S. Yang, C. Yang, Z. Li, J. Wang, and W. Ge, "Poly(N -vinylcarbazole) (PVK) Photoconductivity Enhancement Induced by Doping with CdS Nanocrystals through Chemical Hybridization," *The Journal of Physical Chemistry B*, vol. 104, pp. 11853–11858, Dec. 2000.

[139] F. J. Zhang, Z. Xu, D. W. Zhao, S. L. Zhao, L. W. Wang, and G. C. Yuan, "The effect of DCJTB doping concentration in PVK on the chromatic coordinate of electroluminescence," *Physica Scripta*, vol. 77, p. 055403, May 2008.

[140] K. D. Ley and K. S. Schanze, "Photophysics of metal-organic pi -conjugated polymers," *Coordination Chemistry Reviews*, vol. 171, pp. 287–307, 1998.

[141] F. Scandola and V. Balzani, "Energy-transfer processes of excited states of coordination compounds," *Journal of Chemical Education*, vol. 60, p. 814, Oct. 1983.

[142] B. H. Boo, S. Y. Ryu, H. S. Kang, and S. G. Koh, "Time-resolved Fluorescence Studies of Carbazole and Poly(N-vinylcarbazole for Elucidating Intramolecular Excimer Formation," *Journal of the Korean Physical Society*, vol. 57, p. 406, Aug. 2010.

[143] B. Kumar, M. Llorente, J. Froehlich, T. Dang, A. Sathrum, and C. P. Kubiak, "Photochemical and photoelectrochemical reduction of CO_2.," *Annual review of physical chemistry*, vol. 63, pp. 541–69, Jan. 2012.

[144] E. Portenkirchner, D. Apaydin, G. Aufischer, M. Havlicek, M. White, M. C. Scharber, and N. S. Sariciftci, "Photoinduced Energy Transfer from Poly(N-vinylcarbazole) to Tricarbonylchloro-(2,2'-bipyridyl)rhenium(I).," *ChemPhysChem*, pp. 1–6, Aug. 2014.

[145] F. Garnier, G. Tourillon, M. Gazard, and J. Dubois, "Organic conducting polymers derived from substituted thiophenes as electrochromic material," *Journal of Electroanalytical Chemistry and Interfacial Electrochemistry*, vol. 148, pp. 299–303, June 1983.

[146] C. R. Mason, P. J. Skabara, D. Cupertino, J. Schofield, F. Meghdadi, B. Ebner, and N. S. Sariciftci, "Synthesis and properties of end-capped sexithiophenes incorporating the ethylene dithiothiophene unit," *Journal of Materials Chemistry*, vol. 15, no. 14, p. 1446, 2005.

[147] A. Mozer, N. Sariciftci, A. Pivrikas, R. Österbacka, G. Juška, L. Brassat, and H. Bässler, "Charge carrier mobility in regioregular poly(3-hexylthiophene) probed by transient conductivity techniques: A comparative study," *Physical Review B*, vol. 71, p. 035214, Jan. 2005.

[148] H. Gerischer and C. W. Tobias, eds., *Advances in Electrochemical Science and Engineering*, vol. 2 of *Advances in Electrochemical Sciences and Engineering*. Weinheim, Germany: Wiley-VCH Verlag GmbH, Dec. 1991.

[149] P. S. Sharma, A. Pietrzyk-Le, F. D'Souza, and W. Kutner, "Electrochemically synthesized polymers in molecular imprinting for chemical sensing.," *Analytical and bioanalytical chemistry*, vol. 402, pp. 3177–204, Apr. 2012.

[150] K. Tanaka, T. Shichiri, S. Wang, and T. Yamabe, "A study of the electropolymerization of thiophene," *Synthetic Metals*, vol. 24, pp. 203–215, May 1988.

[151] R. J. Waltman, J. Bargon, and A. F. Diaz, "Electrochemical studies of some conducting polythiophene films," *The Journal of Physical Chemistry*, vol. 87, pp. 1459–1463, Apr. 1983.

[152] M.-a. Sato, S. Tanaka, and K. Kaeriyama, "Electrochemical preparation of highly conducting polythiophene films," *Journal of the Chemical Society, Chemical Communications*, no. 11, p. 713, 1985.

[153] X. Wang, G. Shi, and Y. Liang, "Low potential electropolymerization of thiophene at a copper oxide electrode," *Electrochemistry Communications*, vol. 1, pp. 536–539, Nov. 1999.

[154] "Abundance of Elements in the Earth's Crust and in the Sea," in *CRC Handbook of Chemistry and Physics* (D. R. Lide, ed.), pp. 14–17, Boca Raton, Florida: CRC Press/Taylor and Francis, 90th editi ed., 2009.

[155] T. Kelly and G. Matos, "U.S. Geological Survey, Rhenium Mineral commodity statistics," tech. rep., Historical statistics for mineral and material commodities in the United States: U.S. Geological Survey Data Series 140, 2012.

[156] J. M. Smieja, M. D. Sampson, K. a. Grice, E. E. Benson, J. D. Froehlich, and C. P. Kubiak, "Manganese as a substitute for rhenium in CO_2 reduction catalysts: the importance of acids.," *Inorganic chemistry*, vol. 52, pp. 2484–91, Mar. 2013.

[157] K. A. Grice and C. P. Kubiak, *Recent Studies of Rhenium and Manganese Bipyridine Carbonyl Catalysts for the Electrochemical Reduction of CO_2*, vol. 66. Elsevier Inc., 1 ed., 2014.

[158] M. Bourrez, F. Molton, S. Chardon-Noblat, and A. Deronzier, "Mn(bipyridyl)(CO)$_3$Br: An Abundant Metal Carbonyl Complex as Efficient Electrocatalyst for CO_2 Reduction.," *Angewandte Chemie (International ed. in English)*, pp. 9903–9906, Sept. 2011.

[159] M. W. Kanan and D. G. Nocera, "In situ formation of an oxygen-evolving catalyst in neutral water containing phosphate and CO_2^+.," *Science (New York, N.Y.)*, vol. 321, pp. 1072–5, Aug. 2008.

[160] M. W. Kanan, Y. Surendranath, and D. G. Nocera, "Cobalt-phosphate oxygen-evolving compound.," *Chemical Society reviews*, vol. 38, pp. 109–14, Jan. 2009.

[161] K. W. Pratt, W. F. Koch, Y. C. Wu, and P. A. Berezansky, "Molality-based primary standards of electrolytic conductivity (IUPAC Technical Report)," *Pure and Applied Chemistry*, vol. 73, no. 11, pp. 1783–1793, 2001.

[162] V. Aguilella, S. Mafe, J. Manzanares, and J. Pellicer, "Current-voltage curves for ion-exchange membranes. Contributions to the total potential drop," *Journal of Membrane Science*, vol. 61, pp. 177–190, Jan. 1991.

[163] J. Krol, "Concentration polarization with monopolar ion exchange membranes: current voltage curves and water dissociation," *Journal of Membrane Science*, vol. 162, pp. 145–154, Sept. 1999.

[164] L. Marder, E. M. Ortega Navarro, V. Pérez-Herranz, A. M. Bernardes, and J. Z. Ferreira, "Evaluation of transition metals transport properties

through a cation-exchange membrane by chronopotentiometry," *Journal of Membrane Science*, vol. 284, pp. 267–275, Nov. 2006.

[165] K. Lackner, "Capture of carbon dioxide from ambient air," *The European Physical Journal Special Topics*, vol. 176, pp. 93–106, Sept. 2009.

[166] J. Kunze and U. Stimming, "Electrochemical versus heat-engine energy technology: a tribute to Wilhelm Ostwald's visionary statements.," *Angewandte Chemie (International ed. in English)*, vol. 48, pp. 9230–7, Jan. 2009.

[167] W. Ostwald, *Kunst und Wissenschaft*. Leipzig: Verlag von Veit & Comp., 1905.

Index

ab initio, 64
absorbance spectra, 58
alkylammonium salt, 126
aprotic solvents, 73
Arrhenius equation, 111
aryl-conjugated, 126
asymmetric stretching, 125
atmospheric carbon dioxide, 10
ATR-FTIR measurements, 42

back-bonding (back donation), 53
band bending, 134
Beer-Lambert law, 54
bidentate ligands, 53
bimolecular reaction rate, 149
Brownian motion, 27

capacitance, 26
carrier concentration (N_A), 135
cation exchange membrane, 173
charge transfer, 18, 131
chelating imino groups, 83
CO_2 capture, 160
cobalt–phosphate catalyst, 172
conjugated polymer, 154
coordination number, 49
coordination site, 85
coordinative interactions, 91
coulomb attraction, 18, 131
critical scan rate, 25
crystal field splitting, 52
current energy supply, 9

cutoff filter, 108
cyclic voltammetry, 27

d-d transitions, 54
d-orbital, 66
d-orbitals, 52
degree of freedom, 62
density functional theory, 61
depletion width (W), 133
Dexter resonance energy transfer, 149
diffusion coefficient, 80
diffusion layer, 26
diffusion limited current density, 139
dimer species, 80
donor-acceptor, 145
doping level (donor density), 135
DTGS detector, 38
dual luminescence, 60
dynamic viscosity, 23

electric double layer, 27
electrical double layer, 25
electro-polymerization, 117
electrochemical activation, 14
electrode surface concentration, 24
electrode surface layer, 75
electrolyte conductivity, 173
electron transfer, 111
electron transfer rate, 22
electron-hole pair, 18, 131
electropolymerisation, 34
eluent source, 46

INDEX

energy gap, 136
energy transfer, 150
excitation spectra, 145
excited state, 109
excited state lifetime, 149
exciton, 18, 132

Förster resonance energy transfer, 149
Faradaic efficiencies, 90
Faradaic efficiency, 32
Fermi level, 132
ferrocene, 27
Fick's laws, 22
Fick's second law of diffusion, 30
field dissociation, 27
flat band potential, 133
flat-band potential, 135
fluorescence lifetime, 149
Fourier's law, 23
Franck-Condon-Principle, 109
FTIR measurements, 38

gas chromatography , 44
Gaussian09, 61
Gibbs energy of reaction, 174
glassy carbon electrode, 73

H-cell, 29
Hartree-Fock, 64
headspace, 31
Helmholtz model, 27
Henry constant, 33
Henry's Law, 33
heterocyclic aromatic system, 83
heterogeneous catalysis, 113
hexafluorophosphate, 27
HOMO, 64
homogeneous catalysis, 70
hydrodynamic radius, 23

imidazole, 92
inorganic semiconductor, 17, 131
intraligand emission, 60
ion chromatography, 46
isomers, 50

Jablonski-Diagramm, 54
junction potential, 172

Levich equation, 80
light scattering, 146
low pass, 108
low-spin-configuration, 52
LUMO, 64

manganese, 169, 170
Marcus theory, 110
MCT detector, 38
membrane, 172
metal-to-ligand charge transfer, 55
mid infrared, 38
MLCT, 57
molecular orbital, 64
monochromator, 37
Mott-Schottky analysis, 135
multichromophore, 60

Nernst equation, 25

Octahedral point group, 49
ohmic losses, 172
ohmic potential drop, 172
oil production, 11
oligomer radicals, 157
one-compartment cell, 28
organic semiconductor, 17, 131
outer Helmholtz Plane, 26
oxide layer, 128
oxygen evolution, 172

p-GaP semiconductor, 17
p-type semiconductor, 133
para substituted, 79
parity forbidden, 54
partial pressure, 174
photo catalyst, 105
photochemical system, 109
photocurrent, 139
photoluminescence, 37, 58
photophysical data, 59
photoresponse, 140
photosensitizer, 107
photostability, 52
photovoltaic current densities, 170
poly(3-hexylthiophene), 137
polymer backbone, 154
polymerization, 113
polypyrrole matrix, 114
polythiophene, 153
potentiodynamic, 155
potentiodynamic film formation, 114
pyridinium, 91
pyridinium radical, 91

quantum mechanics, 61
quantum yield, 105
quartz cuvette, 29
quasi reference electrode, 28
quasi-reversible, 73
quencher, 148
quenching constant, 149
Quinacridone, 160

radical addition, 127
Randles-Sevcik equation, 95
Randles–Sevcik equation, 84
rate constant, 80
rectification, 134

redox potentials, 14
reorganization energy, 110
rhenium metal pellet, 169
RMS roughness, 127
rotating disk, 80

schematic mechanism, 71
Schottky, 18, 131
semiconductor, 132
silicon diode, 146
solid-liquid interface, 145
solvatochromism, 53
Sonogashira coupling, 71
space charge, 134
specific conductivity, 173
Stern model, 27
Stern-Volmer constant, 149
Stern-Volmer Plot, 148
Stokes-Einstein equation, 23
surface coverage, 123
symmetry elements, 49
synthetic chemical fuels, 170

temperature effect, 139
tetrabutylammonium, 27
thermocouple, 141
time constant, 141
trace crossing, 80
transfer coefficient, 21
transition metals, 69
transmission cell, 38
transmittance, 54
triethanolamine, 46, 57
turn over frequency, 122
turn over number, 107

UV-vis absorption, 36

workfunction, 133

Acknowledgment

First and foremost I want to thank Professor Niyazi Serdar Sariciftci who enabled this dissertation, for your support and supervision during the last years.

I am very grateful to Professor Günther Knör for his supervision and support. Your deep knowledge of photochemistry enriched my work and thought me a lot during our discussions. I would like to thank Helmut Neugebauer for his critical comments and discussions that made me think twice about my results and strengthened the arguments a lot. Many thanks to Professor Dieter Meissner for your endless and patient attempts to teach me electrochemistry and thermodynamics during my studies and for continuing to do so ever when we meet again.

I would like to thank Tsukasa Yoshida for the possibility to do exciting research at your group in Yonezawa and for pointing out that being different has value. Special thanks belong to Patchanita Thamyongkit for our good collaboration, providing materials and waiting patiently for the results. Many thanks for your humor and support in all those meetings at LIOS. I would like to thank Sean Shaheen for encouraging me to attend the I-Camp summer school in Boulder and the good time and fruitful learnings I enjoyed there. Furthermore I want to thank Gregor Waldstein for the support within the Solar Fuel project, opening for me the possibility for graduate studies at LIOS.

Special thanks to my coworkers, former and present. Particularly to Jacek Gasiorowski, Kerstin Oppelt and Christoph Ulbricht for each time we shared a cup of coffee and discussed about science and life. *"Stop it!"* To Dogukan Apaydin, Markus Scharber, Matthew White and Stefanie Schlager for long and patient discussions about science. Furthermore to Eric Glowacki, Marek Havlicek, Philipp Stadler, Elisa Tordin, Christina and Sandra Enengl, Gottfried Aufischer, Elham Kianfar, Mateusz Bednoroz, Martin Egginger, Anita Fuchsbauer, Sandra Kogler, Martin Kruijen, Gebhard Matt, Alberto Montaigne-Ramil, Almantas Pivrikas, Hans-Jürgen Prall, Doris Sinwel, Peter Trefflinger and especially to all those who are not mentioned here but should be.

Gabriele Hinterberger deserves special thanks for many patient measurements that I could not find time to do myself. I would like to thank Nadja Danklmaier and Gerda Kalab for many things, especially however for giving counsel on problems far beyond issues solely concerning the work. Furthermore I would like to thank Birgit Paulik and Petra Neumaier for all the support in administrative work along the way.

Last but not least I want to thank my family. Especially my parents Hermine and Karl and my brothers, Karl and Herbert Portenkirchner for all their support in a long and laborious journey. For providing a place where I am always welcomed, where I can rest and regenerate my strength for the next steps. Siegfried Rettenegger for giving compassionate advice in the most difficult moments. Silvia and Peter Geistlinger for your generosity and warmth that encouraged me in many ways, Anika for reminding me that fairies, wizards and mermaids do exist and for putting everything in perspective, and of course to you Katharina, for your love and patience. You are the sunshine in my life!

i want morebooks!

Buy your books fast and straightforward online - at one of the world's fastest growing online book stores! Environmentally sound due to Print-on-Demand technologies.

Buy your books online at
www.get-morebooks.com

Kaufen Sie Ihre Bücher schnell und unkompliziert online – auf einer der am schnellsten wachsenden Buchhandelsplattformen weltweit!
Dank Print-On-Demand umwelt- und ressourcenschonend produziert.

Bücher schneller online kaufen
www.morebooks.de

OmniScriptum Marketing DEU GmbH
Heinrich-Böcking-Str. 6-8
D - 66121 Saarbrücken
Telefax: +49 681 93 81 567-9

info@omniscriptum.de
www.omniscriptum.de

Printed by Books on Demand GmbH, Norderstedt / Germany